Great Guitar Tone with IK Multimedia AmpliTube®: The Official Guide

Michael Ross

Course Technology PTR
A part of Cengage Learning

COURSE TECHNOLOGY
CENGAGE Learning™

Australia • Brazil • Japan • Korea • Mexico • Singapore • Spain • United Kingdom • United States

COURSE TECHNOLOGY
CENGAGE Learning™

Great Guitar Tone with IK Multimedia AmpliTube®: The Official Guide
Michael Ross

Publisher and General Manager, Course Technology PTR: Stacy L. Hiquet

Associate Director of Marketing: Sarah Panella

Manager of Editorial Services: Heather Talbot

Marketing Manager: Mark Hughes

Executive Editor: Mark Garvey

Project Editor/Copy Editor: Cathleen D. Small

Interior Layout Tech: MPS Limited, a Macmillan Company

Cover Designer: Luke Fletcher

Indexer: Valerie Haynes Perry

Proofreader: Chuck Hutchinson

For product information and technology assistance, contact us at
Cengage Learning Customer & Sales Support, 1-800-354-9706

For permission to use material from this text or product, submit all requests online at **www.cengage.com/permissions**
Further permissions questions can be emailed to
permissionrequest@cengage.com

Library of Congress Control Number: 2011920247

ISBN-13: 978-1-4354-5842-0

ISBN-10: 1-4354-5842-7

Course Technology, a part of Cengage Learning
20 Channel Center Street
Boston, MA 02210
USA

Cengage Learning is a leading provider of customized learning solutions with office locations around the globe, including Singapore, the United Kingdom, Australia, Mexico, Brazil, and Japan. Locate your local office at: **international.cengage.com/region**

Cengage Learning products are represented in Canada by Nelson Education, Ltd.

For your lifelong learning solutions, visit **courseptr.com**

Visit our corporate website at **cengage.com**

Printed in the United States of America
1 2 3 4 5 6 7 13 12 11

For my wife, Elizabeth, with her musician's ear and her editor's eye.

Acknowledgments

Many thanks to Mark Garvey at Course Technology PTR and to my project editor, Cathleen Small. Thanks, too, to the folks at IK Multimedia—Gary Kerzner, Daniel Boatman, and Starr Ackerman—for all of their help and for making such a cool product.

About the Author

Michael Ross is a musician/writer/producer living in New York City. He has toured the United States, as well as Cuba, Norway, Puerto Rico, and Iceland, playing everything from blues and country to avant-pop and electronica.

Michael has authored two books: *Getting Great Guitar Sounds* (Hal Leonard, 1998) and *All About Effects* (Hal Leonard). He contributes to *Guitar Player* magazine, *Premier Guitar*, *Guitar Edge*, *EQ*, *Electronic Musician*, *In Tune*, *Sound On Sound*, and Gearwire.com.

As a producer, Michael has recorded artists in San Francisco and New York. He has also taught guitar and lectured at the Blue Bear School of Music in San Francisco.

Contents

Chapter 6 The Cabinet 63

Chapter 7 Miking 73

Preface

Welcome to *Great Guitar Tone with IK Multimedia AmpliTube: The Official Guide*. Technical facility on any instrument is a must if you are to fully express yourself. But being able to make your fingers fly on your guitar is only half the equation—maybe less. Listeners who aren't musicians, and even many who are, respond first to the sound that you are putting out. Would Wes Montgomery, Charlie Christian, Jimmy Hendrix, Eddie Van Halen, and Stevie Ray Vaughan have reached their legendary status had their tone been weak, tinny, and anemic? I think not.

It wasn't much more than a decade ago that getting great guitar tone on a track required thousands of dollars worth of equipment—owned or rented. It also meant hundreds or thousands of dollars of studio time while the engineer and/or his assistant tried different microphones, moving them an inch at a time to find the sweet spot in front of the speaker.

Once recorded, the sound was set. If it didn't seem appropriate after later parts were added or the song was being mixed, it was scrapped, and a new part meant starting the whole expensive, time-consuming process over.

Getting a good take meant using up precious tracks with repeated attempts or, if only one track was available, erasing what might have been a keeper because you thought you could do it better.

In the digital age, ending up with great-sounding guitar tracks on a recording is infinitely easier. Some recording software offers playlists on each track, where you can record multiple takes without moving to another track. Then you can choose the playlist you want or pick the parts you like from each playlist and comp them together. Even on DAWs lacking this feature, tracks are plentiful, limited largely by your computer's CPU capacity. With the rise of home studios, recording costs have become minimal, making studio time much more affordable.

Then there is modeling software, offering a choice of vintage and modern amplifiers and a wealth of effects for the price of a single boutique stompbox. In addition to presenting a cornucopia of available tones, modeling software means that you can change the sound of the take at any point in the process—right through the final mix!

In an ideal world, many guitar players would still prefer access to dozens of real amps and pedals, combined with endless hours in a professional studio to experiment. Unfortunately, most guitarists don't live in that world. I have used AmpliTube since its introduction and have found it delivers a sound and feel that makes playing through modeling software less like a compromise and more like an adventure. I employ AmpliTube on my own recordings and on professional sessions for others and have reveled in the tonal options and the convenience.

Making music is about grabbing the inspiration, and nothing can kill the initial excitement of a new idea like sitting around waiting for an amplifier to be miked. With AmpliTube, I can switch tones on a dime and quickly follow my inspiration where it leads me.

Speaking of tones, I think we can all agree that what constitutes great guitar tone is, on a certain level, up for grabs. Country fans and Nirvana fans might disagree on which tones are great (though these days not as much as you might think).

Through the years there have been many guitarists known as much for their tone as for their playing ability: Eddie Van Halen, Eric Clapton, Kurt Cobain, Dave Gilmour, Jack White, the Edge, Robben Ford, Tom Morello, and Stevie Ray Vaughan come to mind immediately. I am sure that you can think of others or may disagree with these, but their wide range of sounds proves my point: Great tone is, in some sense, in the ear of the beholder. In another sense, all of these artists represent some of the same things with regard to tone: They are all in total control of their sound, and it is a sound that inspires them to perform at their peak, thus inspiring millions of listeners as well. These two things—control and inspiration—are what I hope to help you with here. By learning how to control the numerous digital models of analog sounds produced by AmpliTube, you will be able to easily access the inspired sounds in your head and ultimately inspire others with them.

Introduction

When IK Multimedia launched AmpliTube at the beginning of the millennium, it represented a godsend to computer-based recordists working with digital audio workstations (DAWs). Line 6's Amp Farm had introduced guitarists to the advantages of software-modeled amps but was only available to hardware-enhanced Pro Tools TDM users. AmpliTube let those who worked in all of the other DAWs enjoy the convenience and versatility of having a variety of amplifier sounds available at the touch of a mouse. A decade later, it remains one of the most popular and best-sounding examples of digital amplifier and effects modeling.

At this point, AmpliTube 3 represents the culmination of this modeling software (though at some point there will no doubt be an AmpliTube 4, and so on). AmpliTube 3 incorporates all of the amps and effects from AmpliTube 2, as well as those found in AmpliTube Metal and AmpliTube Jimi Hendrix. It offers 51 stompbox effects, 31 guitar and bass amp modules, 46 cabinet emulations, 15 microphone models, and 17 rack effects. If you're a fan of Fender amps, I highly recommend that you add the AmpliTube Fender package that concentrates on 12 Fender amps, rendering them with uncanny accuracy. Likewise, if you do a lot of bass recording, you will want to add the Ampeg SVX package so you can access amps such as the classic SVT and B-15R models. AmpliTube 3's open architecture allows these packages to integrate seamlessly with AmpliTube 3, no longer requiring X-Gear for access.

As of this writing, IK Multimedia has plans to start a Custom Shop where you can purchase AmpliTube on an a la carte basis. It will offer packages that represent specific amplifier brands, such as AmpliTube Fender, as well as inexpensive modular packages of individual effects and amps. For the purposes of this book, we will employ AmpliTube 3.

In standalone mode, AmpliTube 3 offers everything you need to begin honing your recorded and live guitar tone. Its 4-track audio player/recorder with Speed Trainer will allow you to practice recording and overdubbing your guitar, even if you do not yet have any recording software in your computer. It will allow you to use one instance of AmpliTube 3 per track, much the same way as you would use it as a plug-in in a recording studio. Its digital metronome will help you keep your licks in time, while its panning capabilities will help you learn to combine and place your parts in the stereo spectrum.

Getting great tone when recording your guitar is achieved the same way you achieve fluency on your instrument—through practice. Just as you practice scales, modes, chords, riffs, and runs, AmpliTube 3 will let you practice using effects, amps, and microphones as you go through this book to constantly improve your sound.

Figure I.1 Roughly $1,000 worth of pedals.

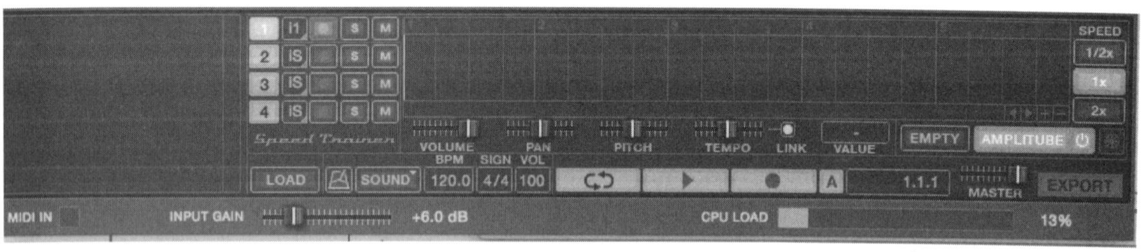

Figure I.2 AmpliTube's 4-track recorder.

The Technology

AmpliTube's sound and feel has been excellent from the first version. Still, as with all digital technology, further research and increasingly more powerful processing in each successive generation of computers has allowed the IK engineers to constantly refine the detail of the sonic replication produced by their amp and effect modeling. AmpliTube 3 reflects this, having ultra-accurately "remastered" all the AmpliTube sounds with their new third-generation DSM (Dynamic Saturation Modeling) technology.

The Italian company spares no time or expense in their efforts to create the most accurate gear models possible. All of the original amps and effects are researched and collected—yes, they actually own hardware editions of all of the original vintage amps modeled in AmpliTube.

As someone who worked in a vintage gear shop for years, I can tell you that all versions of the same year and model amplifier are not created equal. IK takes particular care to find the best-sounding example of a particular model, with era-appropriate speakers. The original amplifier or effect is measured, and a virtual representation of its circuit is coded in software, matching each component's value as well as circuit details. Then they compare the original amplifier or effect to the modeled version. Ultimately, a series of tweaks fine-tunes the circuit and the sonic characteristics of the model. It can take an average of eight weeks to create single model of amp or effect.

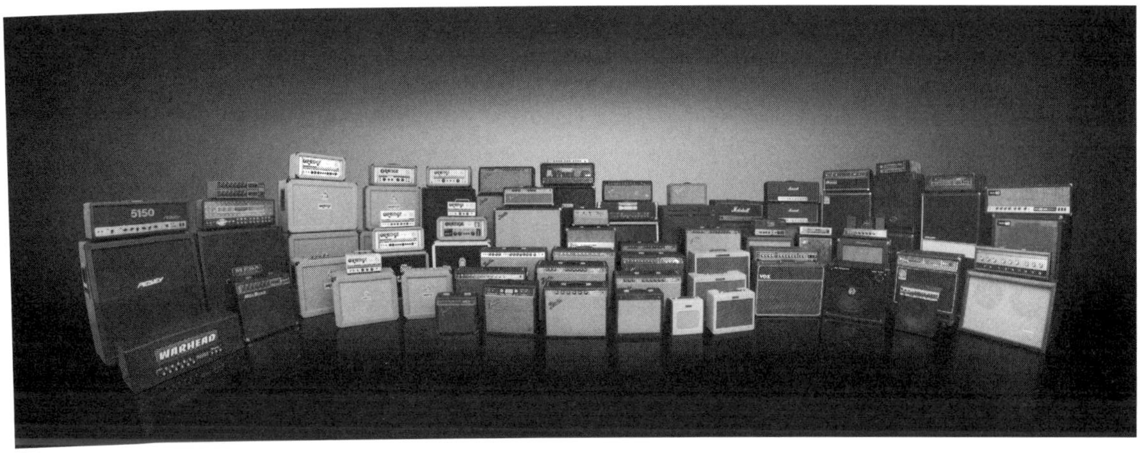

Figure I.3 IK Multimedia's amplifier collection.

With AmpliTube 3, IK's new VRM (Volumetric Response Modeling) technology allows you to double-mike the amp cabinet with a pair of movable microphones. The same technology also provides a highly accurate rendering of the complex sound produced by a rotary speaker. Impulse-based reverbs are now used in the entire chain, from spring reverb to room ambience.

Installation

Whether you have purchased the box version or the download version of AmpliTube 3, installation is simple—once you know where to find everything.

- Macintosh users can just double-click on the Install icon and follow the onscreen instructions. The standalone version is installed in the Applications folder. The documentation is placed in Library\Documentation\IK Multimedia.
- Windows users must double-click on the Install icon and follow the onscreen instructions. The installer will ask you to confirm the destination folder, where you will also find the documentation. To launch the standalone version, go to Start > All Programs. The default destination folder is Program Files\IK Multimedia. AmpliTube 3 should then also appear as a plug-in in whichever DAW you use.

Registration

Be sure to register and authorize AmpliTube. Computers crash and can be lost or stolen. Registration and authorization enable permanent unrestricted access and functionality on more than one computer and let you take advantage of all the User Area features. To register and authorize, just follow the instructions provided in the Product Authorization Wizard, a smart application integrated to the IK website that guides you through the process. The Wizard will pop up when you launch AmpliTube 3 for the first time. Information about AmpliTube 3 installation is available in the Installation and Authorization Manual, located at:

- Start\Programs\IK Multimedia\AmpliTube 3\Installation and Authorization Manual (for Windows users)
- HD\Library\Documentation\IK Multimedia\AmpliTube (for Mac users)

Proper installation and registration will ensure smooth operation of the software as well as keeping you apprised of updates and offerings. Keeping your software updated to the latest version is crucial to avoiding bugs and issues in this age of constant computer system updates. It will also keep AmpliTube operating at peak efficiency.

Preparation

Though you are no doubt itching to start playing with AmpliTube, there are a few things we must cover first.

- The best software cannot cover up the wrong choice of instrument or one that plays badly and sounds terrible through any amp. So in Chapter 1, "The Guitar," we will cover the guitar itself.
- Before you can start playing your instrument through AmpliTube, you need to get its sound into the computer. Chapter 2, "The Audio Interface," will explain your audio interface options and how to use them optimally.
- Then, though you can certainly manipulate AmpliTube with a mouse, we will cover some of the hardware control options in Chapter 3, "Control." These will make using AmpliTube faster and easier and keep the inspiration flowing.
- How you hear the sound coming through the software will greatly affect how others hear it on your recording, so in Chapter 4, "Monitors," we will discuss monitoring the sound coming out of AmpliTube 3.

It's time to enter the world of recording guitar. One thing that I will emphasize throughout this book is that you should experiment. You will read many rules, tips, and tricks, but if you are to develop your own instantly recognizable sound, nothing is more important than experimenting. AmpliTube allows you to put your stompboxes in any order you choose, and it gives you dozens of options for mixing and matching preamps, EQ sections, and power stages of amps. You can then pair them with a plethora of possible speaker cabinets—and that's just the beginning. I encourage you to try out the craziest configurations you can think of. Who knows what you'll come up with? Above all, have fun!

1 The Guitar

Whether you're using software or hardware, the first step in recording guitar is choosing and preparing the guitar. There is an old acronym in the computer world: GIGO. It means Garbage In Garbage Out; that is, what you get out of the computer is only as good as what you put into it. In this case, it means if your instrument doesn't sound good or play well, your tracks will suffer—no matter how good the sounds you get out of AmpliTube 3. For a full course in understanding how to optimize your guitar sound, allow me to immodestly recommend my book: *Getting Great Guitar Sounds* (Hal Leonard, 1998). I will summarize some of those tips here.

Choosing the Instrument

As long as we are trotting out adages, consider this one: "the right tool for the right job." Deciding which guitar (or bass) to use for your recording might be as simple as using the only one that you own and play—that is your sound, and you're sticking to it. Other situations may call for some decision making.

One of the joys of recording is the wealth of possible sounds available. AmpliTube will provide you with the vast majority of those sounds, but you might be surprised by the range of tones available from the instrument itself, before you ever plug it in. The type of guitar, the strings, the picks, where you pick along the string—all of these factor into the sound going into the computer, and thus the sound coming out.

Let's assume you have access—either through your own collection, friends, or rental—to a wide range of instruments. The first, perhaps obvious, choice is whether you are laying down a guitar or a bass part. Or, maybe the part needs to be a baritone guitar (a slightly longer-scale instrument tuned down to A). After you have made this choice, you need to pick a solid body, semi-hollow body, or fully hollow body instrument—the more hollow the body, the more "woody" the sound.

The next options are single-coil pickups or humbuckers. It is important to make a distinction here between "hum-canceling" pickups and humbuckers.

- *Humbucker* generally refers to pickups containing two coils side by side, wired in series and reverse polarity to eliminate 60-cycle hum, such as the ones on a Gibson

Figure 1.1 A quality instrument like this Paul Reed Smith will help you achieve a quality sound.

Les Paul. This type of pickup typically has a more powerful output than a single coil and generally has a darker sound.

- *Hum canceling* refers to single-coil pickups that have been designed to eliminate the hum associated with the vintage variety, usually found on Fender Stratocasters or Telecasters. A humbucker pickup is always hum canceling, but, to avoid confusion, a single-coil pickup is not normally called a *humbucker*, even if it cancels hum.
- P-90 style pickups are another option. A P-90 is a powerful single-coil pickup with a throaty midrange sound.

Computer screens can cause guitars to hum and buzz, much like fluorescent lighting. A properly shielded instrument and shifting your position can make recording with vintage single-coils possible, but if you are going to be doing much recording in front of a computer and opt for a single-coil sound, you may want to invest in some hum-canceling single-coils.

Figure 1.2 A humbucker pickup (second from left) is essentially two single-coils joined together.

Preparing the Instrument

No amount of spectacular playing or great amp and effect tones will make up for an instrument that is poorly intonated or has dead strings and buzzing frets. It is vital that any instrument you use for recording (or playing live) be in tiptop shape. Here is a checklist for you or your repairperson to make sure that your ideas go down as conceived.

- **Strings.** For starters, change them. New strings will bring out the highs in your instrument. Even if you like a duller sound, remember that highs can be rolled off on the guitar or amplifier or in mixing, but they cannot be added if they aren't there to begin with. Also, old strings will not intonate properly. The gauge of string will affect the sound (lighter will be brighter; heavier will be darker) and the setup, so choose the gauge before you have the instrument set up. Be sure that you keep your strings fresh during the session. If you try to punch in a part played with new strings later in the session when the strings are old, you will not be able to match the sound. How often you need to change them depends on how much you sweat when you play and the corrosive quality of your perspiration; some players can go days, and some need to change their strings every few hours.
- **Setup.** This involves various operations and, unless you know what you're doing, is best left to an expert. A good repairperson will set the height of the strings—called the *action*—to your liking by adjusting the bridge and the truss rod. He or she will also file out the nut slots to prevent the strings from catching or cut a new nut if the old one is contributing to buzzing.
- **Frets.** A setup might also include filing the frets level or even replacing them. Bad frets may buzz, whereas flat frets can prevent proper tuning.

Getting in Tune, Staying in Tune, Playing in Tune

Out-of-tune instruments are a no-no in the world of professional recording. (*Professional* means that you want to get paid at some point.) Here are some tips for getting, staying, and playing in tune.

Figure 1.3 A professional repairman can help your guitar get the best recorded sound.

- **Getting in tune.** Getting your instrument in tune, to the point where chords ring true and all the notes are correct, requires more than a tuner. AmpliTube supplies an excellent chromatic tuner that will detect the note you play and tell you whether it is in tune (see Figure 1.4). But it can't do its job if the instrument is not properly intonated or has bad frets. Proper intonation involves moving the saddles of the bridge forward and back until the octave at the twelfth fret matches the played note at the same fret. You can probably get it close with the AmpliTube tuner, but I recommend that you have a professional do it with a strobe tuner. Remember, this should always be done with a fresh set of strings that are the same gauge as the ones you'll be using. Also, it helps for everyone to share the same tuner in the studio. Different tuners may be calibrated slightly differently, and even though yours says you are "in," you may still be "out" with the others.
- **Staying in tune.** Once you have your instrument set up and intonated, be sure to stretch the strings. This doesn't mean pulling them once, retuning, and you are finished. This means gently but firmly pulling each string, retuning, and repeating the process until they stop going flat. Failing to do so will mean that the string windings around the tuners will continue to tighten and the strings will go flat, especially when bending notes. Speaking of string windings, three or four is plenty for the higher strings, while two may suffice for the low A and E.
- **Playing in tune.** It may seem simple to play the guitar in tune once it is set up and properly tuned. Just put your fingers at the correct frets, right? Although this is

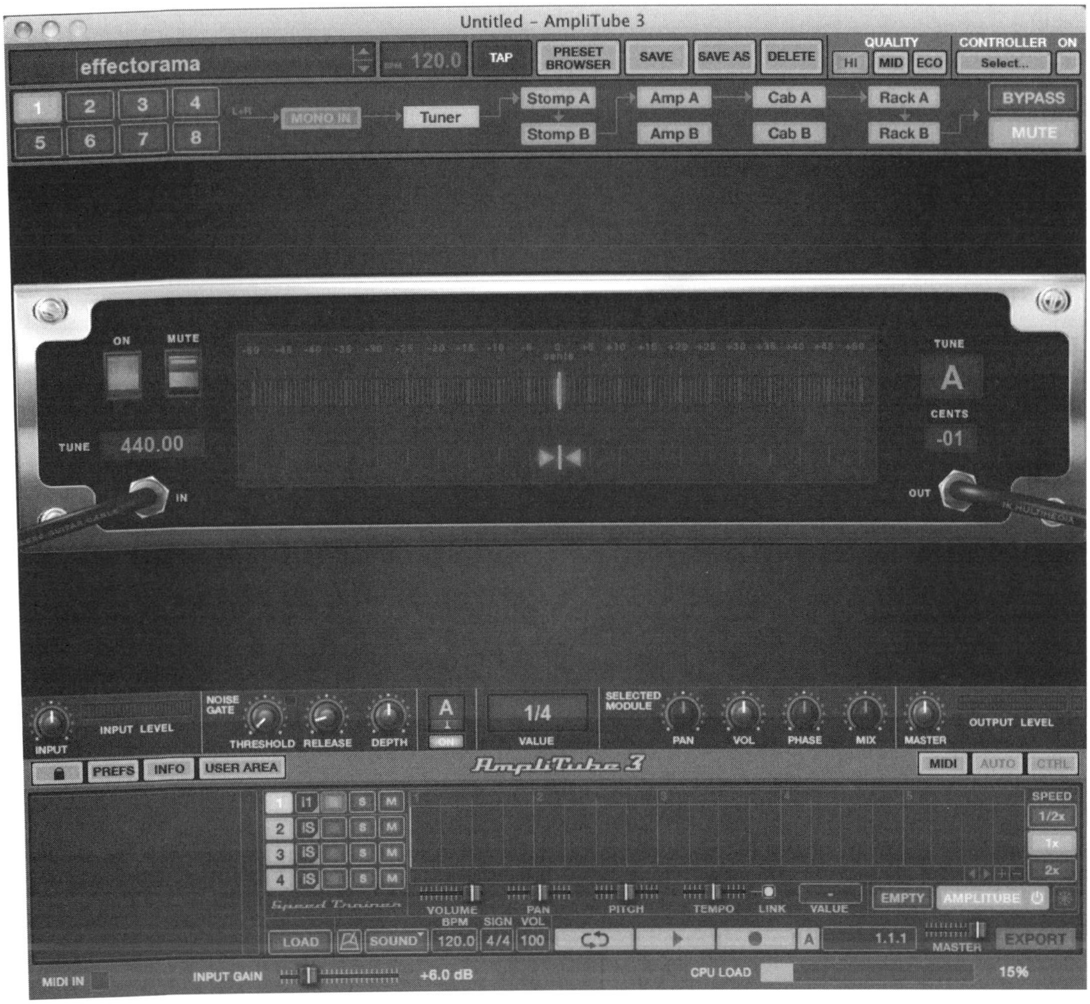

Figure 1.4 AmpliTube's chromatic tuner can help you get and stay in tune.

often sufficient, in some cases there is more to it when you want to perfect your tuning. If your guitar has high frets and you push down too hard or unevenly when playing chords, your playing will sound out of tune. This is often a problem for players who primarily play heavier-stringed acoustic guitar or bass and then attempt to play lighter-stringed electric instruments. It requires only enough pressure to ensure that the string makes solid contact with the fret for the note to ring true—any additional pressure can push the note sharp.

Now that you have your instruments in tiptop shape (and you know the secret to playing in tune), you're ready to record some killer tracks.

Summary Questions

You can find the answers to the Summary Questions in the Appendix at the end of this book.

1. Which has a "woodier" tone: a solid-body guitar or a hollow-body guitar?
2. What is the difference between a hum-canceling pickup and a humbucker pickup?
3. What are two reasons to use new strings when recording?
4. What is the term for the height of the strings off the fretboard?
5. What is the difference between getting in tune and staying in tune?
6. What should you do to your strings to ensure that they stay in tune when they are played and bent?

2 The Audio Interface

At this point you should have AmpliTube installed in your computer. It is now time to choose the method by which you will get the sound from your guitar pickups into your tower or laptop. The analog signal sent from your guitar pickups must be converted to digital information to be processed by AmpliTube. The resulting processed sound must then be converted from digital back to analog audio and sent through speakers or headphones. The device that performs this operation is known as an *audio interface.*

In the previous chapter, I introduced the term *GIGO*, and this applies to interfaces as well. The higher quality interface you use to transmit the sound of your guitar and its pickups to the AmpliTube software, the higher quality sound you will ultimately achieve. The interface you choose will be determined by your needs and your pocket-book—for slamming down riff and song ideas on your laptop while traveling, you probably won't want to carry a full 19″ rack space, $1,600 RME Fireface interface.

You can input your guitar for the cost of a 1/4″ phone jack to 1/8″ mini adapter, spend thousands of dollars on preamps and advanced digital converters, or find a multitude of options in between. Let me stress that the better the interface, the better tone Ampli-Tube will be able to provide. For any kind of professional use, it should offer a minimum of 24-bit A/D and D/A conversion and 44.1/48-kHz operation. (Virtually all but direct interfaces provide this minimum.) The main difference between the cheaper and the more expensive models is the quality of the converters. Plugging directly in is not recommended, as it will not provide AmpliTube with the type of signal that it needs to perform properly.

We will get into the various types of audio interface shortly. First we need to cover a central issue in recording using software modeling: latency.

Latency

Latency refers to the delay between inputting a signal to the computer and hearing the results coming out of the speakers. For guitarists, this means the delay between your pick or finger plucking the strings and you or the audience hearing the sound coming out of the speaker. Latency used to be one of the prime factors that kept guitarists from enjoying the benefits of software modeling. The lag between playing and hearing, often approaching a quarter of a second or more, would drive them crazy. With the availability of faster, more powerful computers and more sophisticated audio interfaces, this delay time can be reduced to a point that is inaudible and, just as important, barely felt—if at all.

There are still latency issues to be dealt with, however, and to minimize latency you need to understand a few terms.

- **CPU.** This refers to the central processing unit of your computer. This is the part that handles the storage and retrieval of all the data and executes program instructions. The CPU power of computers has been increasing enormously in recent years due to factors such as multiple processors: dual-, quad-, and even

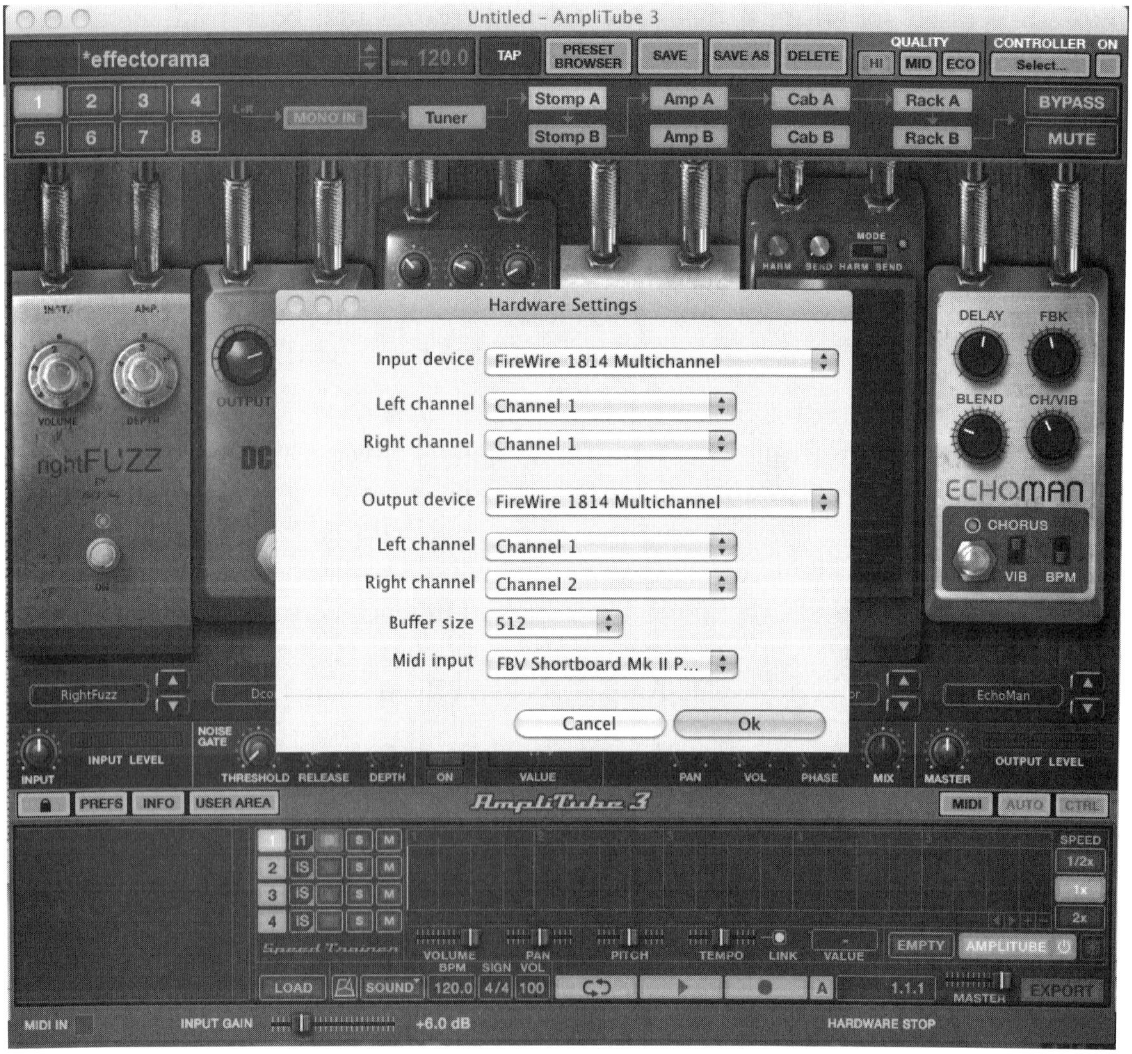

Figure 2.1 In standalone mode, you'll find the buffer settings under Settings > Audio MIDI Setup.

eight-core processors have become common. The size of your computer's CPU will play a part in determining its processing speed, which will in turn affect latency issues.

- **RAM, or random access memory.** This is the part of the computer that allows it to find any data it needs no matter where it is stored. The amount of RAM that your computer has can affect how well AmpliTube works and the latency of the audio.
- **Buffer.** Your computer's buffer is where the data is stored when it is retrieved but before it is used. You can set the size of this buffer when using many audio software applications, such as AmpliTube, Pro Tools, Ableton Live, and so on. The higher the buffer setting, the more complicated the processing that can take place before you experience system issues such as dropouts, audible crackling, or even crashing. The tradeoff is speed of operation. At lower buffer settings, processing is much faster, resulting in lower latency.

The standalone version of AmpliTube's Audio MIDI Setup lets you choose the buffer size you wish to work with. If you are using AmpliTube as a plug-in, the host DAW will have a settings feature to choose a universal buffer size for the processing. The optimum size will always be a tradeoff between speed and performance.

If you are using AmpliTube in standalone mode, employing just a single amplifier model and no effects, you may be able to lower the buffer to as little as 128. But once you start adding dual amps and CPU-hungry effects, such as delays and reverbs, you may have to raise it to 256 or even 512 to avoid glitches in the sound processing.

Plug-In Latency

When you use AmpliTube as a plug-in inside a DAW, in addition to the small amount of latency introduced by AmpliTube, there will be input and output latency caused by getting the audio into and out of the DAW. You'll handle the buffer settings in the DAW preferences rather than in AmpliTube's.

Different DAWs employ different methods to compensate for this delay. Consult your DAW's manual to find out where to adjust the buffer and how to compensate. In the end, as long as you can play along with the track in time, you can always shift the resulting audio track file to make sure that your guitar tracks are in sync with the rest of the audio.

Dealing with Latency

Though theoretically you can install any number of instances of AmpliTube on as many tracks as you like, holding off any final sonic decisions until the mix, in practice there are limitations. The extent of these limitations depends on the power of your computer. Using multiple instances of AmpliTube may require that you increase the buffer to a point where latency becomes too long for comfortable playing. There are a number of tricks for maximizing CPU efficiency and minimizing latency when using AmpliTube.

Figure 2.2 When you are using AmpliTube as a plug-in, the buffer is set in the DAW.

Quality Control

AmpliTube 3's Quality Mode setting lets you choose high quality (HI), medium quality (MID), and economical quality (ECO). One way to reduce latency when tracking parts is to reduce the quality setting. You might be surprised by how good ECO sounds—even though it disables the CPU-hungry room microphones. This allows a lower buffer setting and the use of more instances of AmpliTube. You can always change the settings to HI when you mix. At that point you will be able to increase the buffer as much as you like, as you will no longer be performing, and latency ceases to be an issue.

You can also reduce the CPU usage by modifying some of the settings in AmpliTube's Preferences. Just click on the PREFS button in the lower-left corner of the screen. The Preferences screen will appear. Here you can deselect all of the oversampling and high resolution options; opt for the digital rather than the "real" CPU-hungry convolution versions of amplifier spring reverb and room ambience, and lower the cabinet resolution. Once again, you might be surprised by how good AmpliTube sounds, even without these things engaged. They, too, can be restored when you mix, for optimum fidelity.

Early Guitars

Recording all your guitar parts as early in the project as possible—in other words, over just bass drums and guide vocal—will mean that valuable CPU power is not yet being used to handle many keyboard, background vocal, and percussion tracks and plenty of

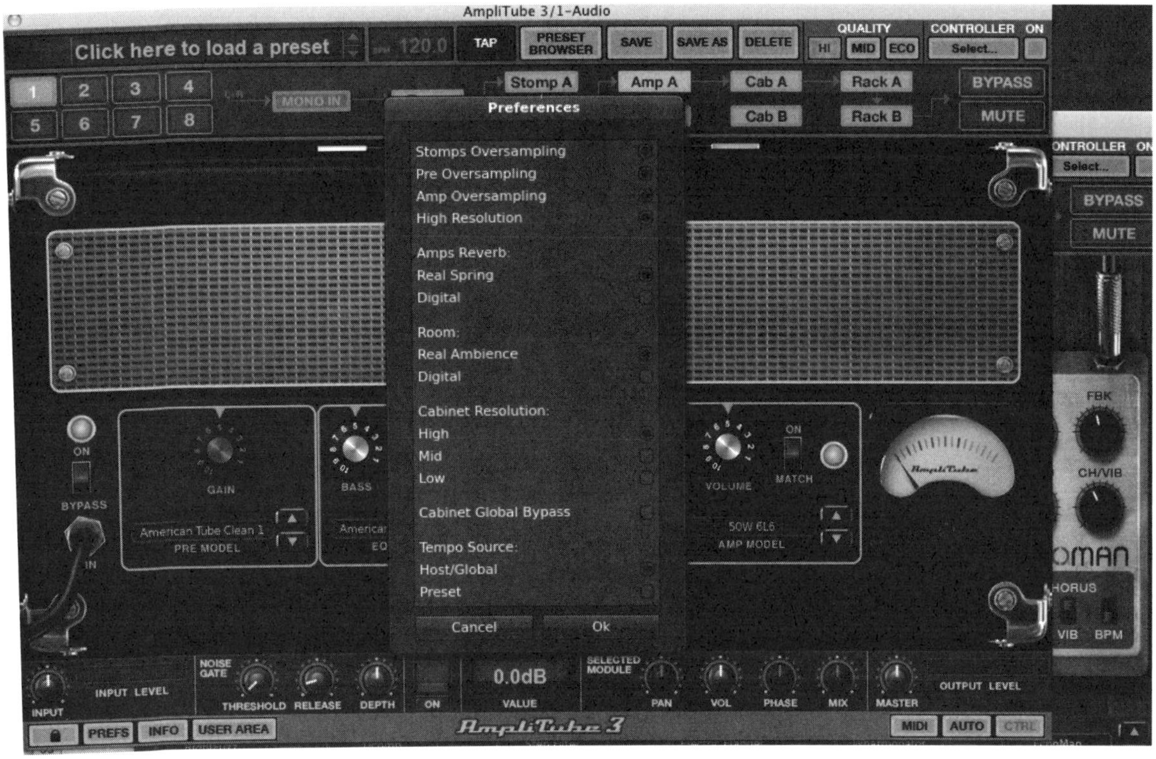

Figure 2.3 The Preferences section lets you disable the high-quality processing to save CPU power.

plug-ins. Once again, this will enable you to keep the buffer settings low and reduce latency. This way, you may be able to use a higher quality mode when recording—better sound can increase inspiration.

Freeze Out

Many DAWs (and AmpliTube's 4 Track Audio Player/Recorder with SpeedTrainer) allow you to "freeze" recorded tracks. This lets them continue to play while reducing the CPU required to handle the audio, automation, and plug-ins on that track. If you are recording guitar tracks late in the session, you may want to freeze as many of the already recorded tracks as possible while you lay down your parts. This should allow you to reduce the buffer size and thus latency.

Radical Solutions

Extreme situations call for extreme measures. For example, you are called for a session to lay down multiple guitar tracks using AmpliTube. It is late in the project, and there are already 30 tracks of audio with many plug-ins and automation involved. You want the highest quality setting of AmpliTube so that you can really hear how your tracks fit into the already complex mix. Here are a couple of radical ideas for getting great tone in trying circumstances.

One solution is to have the whole session rendered as a rough stereo mix and then import that mix into another session. You are now free to add your tracks, each with an instance of AmpliTube set to high quality. Because you are dealing with only one additional stereo audio track—no plug-ins, no automation—you should be able to set the buffer low and experience little latency. The tracks can then be imported into the original session, with your AmpliTube settings intact but still available for later tweaking during mixdown. (Make sure that the DAW you are working in will handle this kind of import.)

If you have AmpliTube installed on your laptop, you can record using the standalone version set to HI, through a separate audio interface, into the main session. Just take a stereo output from the interface that you use with your laptop and send it into the studio's interface. Your laptop's CPU will be dedicated to AmpliTube, allowing you to use complex rigs without requiring any of the session computer's already overworked CPU. This is essentially the same as if you were recording a real amp rig, with the same limitation: no tweaking of settings during the mix. This solution should be largely reserved for situations where the session you are called to do is on a DAW that doesn't have AmpliTube as a plug-in option.

Now that you understand latency, we can talk about the various ways that you can interface with your computer. Keep in mind that, in addition to their superior sound, higher quality interfaces help reduce latency.

What Do You Need?

The kind of interface you choose will reflect your recording needs. Are you looking to merely record ideas and sounds, to be more fully realized later in a professional studio, or do you want to create a finished sellable product? Are you going to be recording at home or on the go? Are you going to be recording only your electric guitar and/or bass, or do you want a microphone input for vocals, acoustic guitar, and percussion?

FireWire versus USB

Audio interfaces send and receive the digital and audio information through either FireWire or USB connectors. Many computers allow both, and both can offer high-quality sound, but there are some advantages and disadvantages to each.

FireWire

FireWire is Apple's name for the IEEE 1394 High-Speed Serial Bus. This original high-speed transfer system is now known as *FireWire 400*, as there is now an even faster version: FireWire 800 (IEEE 1394b). FireWire interfaces are faster than USB interfaces, helping to reduce latency and making them better able to handle multiple track recordings.

FireWire interfaces operate with what is called *peer-to-peer architecture*. Simply put, this means that the audio interface device helps the computer make certain decisions about the audio transfer. USB devices operate on a master-slave system, with the computer as the master—and you-know-who as the slave. This means that the computer must make all the transfer decisions, tying up valuable processing power.

If you are just going to record some guitar parts to see how they fit together, you don't need a FireWire interface. If you plan to do full sessions with many tracks and plug-ins, you will probably want to "fire" up one of these.

Should you choose FireWire, be sure you keep the drivers up to date. Drivers are the software that helps the interface interact with your computer's operating system. As the operating system is upgraded, you may need to upgrade your interface driver. Be sure to check the manufacturing company's website every couple of months to check for new versions that can be downloaded, and definitely check after a system upgrade.

USB

The Universal Serial Bus (USB) is a more common transfer system (thus my iMac has four USB ports to one FireWire port), used in everything from audio transfer devices to cameras and small storage sticks. As mentioned, its master-slave system tends to slow things down a bit. This was more of a problem back in the days of USB 1; with the introduction of USB 2, the gap between FireWire and USB performance has narrowed. USB interfaces can be significantly cheaper. FireWire will still handle more tracks if you are dealing with large sessions and offers lower latency, but for the average guitarist, USB should handle the job. With USB 3 on the horizon, the gap should become even smaller.

Once again, consider your needs and your budget and then go for the best you can find within those parameters.

Some Interface Solutions

There are dozens of audio interfaces available—which has its ups and downs. The good news is that there is doubtless something out there that sounds good and will suit your budget. The bad news is that it is easy for option anxiety to set in. Ultimately, you will have to decide which one is best for you, but I would hate to leave you hanging without offering the benefit of my experience. I haven't tried them all, but here are some of the companies that have proved themselves to me and to other professionals.

Pro Tools

For many years Pro Tools was the professional recording software standard. It is still used throughout the industry, but it is no longer used as exclusively. If you plan to record at home and then bring your tracks to a high-end professional studio, using Pro Tools will allow you to just open up your home sessions on their computer. Keep in mind that more and more top-of-the-line studios are using other DAWs, such as Logic, Cubase, and Nuendo. More people are just importing individual audio tracks into differing DAWs.

As of this writing, Avid, the parent company of Pro Tools, has announced that the current version, Pro Tools 9, will for the first time be functional with non-Avid interfaces. That said, Avid makes a wide range of audio-interface solutions.

Digidesign Digi 003 Factory

The 003 is a full-featured mixing board and interface combo.

Pro Tools HD TDM Systems

If you are ready to mortgage the farm, these processor-assisted systems are the top of the line. The hardware interface part handles much of the processing for the software and allows the use of many more tracks and CPU-greedy plug-ins.

IK Multimedia

IK Multimedia not only makes the AmpliTube software, it also offers a number of interface options.

StealthPlug

The makers of AmpliTube market this simple USB connector that plugs into your guitar at one end and the USB port of your computer at the other end. It's great for whipping out ideas when you're traveling with your laptop.

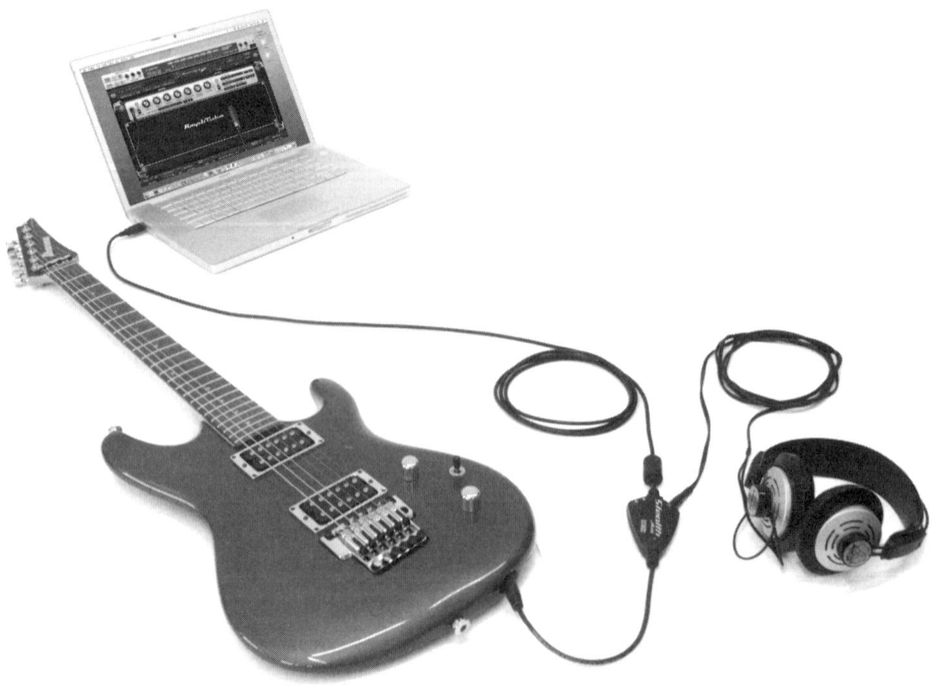

Figure 2.4 The StealthPlug and a pair of headphones are all you need to record on the go.

StealthPedal

This USB unit doubles as a controller for things such as wah, whammy, and volume effects, making it an ideal interface for AmpliTube. In addition, it offers high-quality audio circuits derived from IK's flagship StompIO controller/interface, noise-rejecting inputs, and a sophisticated multiple-power-supply section that allows the StealthPedal to run at internal higher voltages than are usually found in other bus-powered devices. This allows for a much higher grade of component and internal design for better sound quality.

Two separate inputs are provided for Hi-Z or line-level instruments, as well as two separate outputs plus a headphone output. StealthPedal features a software-configurable input gain level and a 109-dBA low-noise input stage.

The fact that it was designed and built by the same people who made the software ensures perfect compatibility. If you plan to make the most of AmpliTube, this interface is your best bet.

Figure 2.5 IK's StealthPedal acts as a high-quality interface and a controller.

M-Audio

Now folded into the Avid group, M-Audio offers a wide range of FireWire and USB interfaces for all purposes and pocketbooks.

Fast Track

M-Audio's Fast Track series—Fast Track, Fast Track Pro, Fast Track Ultra, and Fast Track Ultra 8R—can take the budget-minded player up from entry level to pro studio level. I recommend staying away from the least-expensive models.

MOTU

The manufacturers of the popular Digital Performer DAW also make a number of high-quality interfaces.

MicroBook

Mark of the Unicorn (MOTU) has created this reasonably priced, high-quality, full-featured, pocket-sized interface.

Figure 2.6 Miniaturization has reached the audio interface world with MOTU's MicroBook.

UltraLite-mk3 Hybrid

If you crave more features and additional input and output options, while the unit remains compact, MOTU's UltraLite Hybrid might be for you. Plus, it can be run on USB or FireWire.

ZBox

This is not an interface, but rather a small box with a phone jack input and a built-in phone-jack cable that plugs into an interface. It is designed to mimic the input of a real guitar amp to enhance the reality of the "feel" when you are playing. It works, and I recommend adding it to whichever interface you choose.

Figure 2.7 MOTU's ZBox adds an extra bit of real "feel" when playing through AmpliTube.

Apogee

Apogee's upscale converters are so good that people who have already spent a fortune on a TDM system will often use them instead of the ones in the Pro Tools interface.

Duet

This portable Apogee interface offers both guitar and microphone ins and outs, as well as MIDI ins and outs. It is not cheap, but it sounds great.

Figure 2.8 The Apogee Duet combines small size with big sound.

RME

Depending on the current exchange rate, the German RME products can set you back a pretty penny, but the quality of their converters is widely respected.

Fireface 400

This model and the larger 800 seem to show up wherever high performance is required.

Babyface

A pocket-sized solution for those with deep pockets, the Babyface offers the RME quality (and price) in a Duet/MicroBook-size package.

Direct Monitoring Some audio interfaces advertise direct monitoring as a solution to latency. This means that in addition to being sent through the computer, the guitar signal that you input is sent directly to your headphones or speakers. That direct signal suffers no latency because it never goes through any of the software in the computer. All you will hear, though, is the unprocessed sound of your guitar with none of AmpliTube's effects or amp modeling—not a particularly inspiring way to record. It's better to deal with latency in one of the ways described earlier in this chapter.

The listed interfaces are just a starting point. A quick Internet search will get you detailed specs on all of them, as well as display a plethora of additional manufacturers and models.

Summary Questions

You can find the answers to the Summary Questions in the Appendix at the end of this book.

1. What is an audio interface?
2. What is latency?
3. What is a buffer?
4. Where do I find the buffer settings in AmpliTube?
5. What is AmpliTube's Quality mode?
6. Name three ways that you can deal with latency issues.
7. What are the advantages of USB?
8. What are the advantages of FireWire?
9. What is direct monitoring?

3 Control

The method(s) you use to control AmpliTube will have no effect on the actual tone that you get out of it—regardless of whether you use a mouse or a complex foot controller with four expression pedals, it will not influence the sound at all. However, it will have an effect on your control of that tone.

When recording, you may find yourself in a situation where you are limited to one track for whatever reason. (Yes, it can still happen in this digital age.) Being able to switch sounds mid-track can be difficult, if not impossible, with merely a mouse and a spacebar. If you have all the tracks you need, you can potentially get by with just a mouse—but what about wah-wah and whammy effects?

This is where MIDI control comes in.

MIDI

MIDI stands for *Musical Instrument Digital Interface*. It is the standard protocol adopted by the musical equipment industry to enable synthesizers, drum machines, MIDI controllers, samplers, and computer software to communicate and synchronize with each other. MIDI does not transmit any audio signal. Instead, it sends event messages describing pitch and intensity, as well as signals that adjust parameters such as volume, vibrato, panning, and tempo.

A few terms to understand when dealing with MIDI are:

- **MIDI mapping.** This refers to setting up which parameters of your hardware or software are controlled by which sources on your controllers.
- **CC.** This stands for *continuous controller*. A knob, slider, or expression pedal sends a continuous stream of 127 messages as it is manipulated from one extreme position to the other. A switch can send CC messages as well, usually (but not always) jumping from 0 to 127 to turn something off or on. The CC number helps determine which parameters it controls.

MIDI Controllers

Hardware MIDI controllers come in all shapes and sizes, but they are primarily divided into two types: those controlled by hand and those controlled by foot. For AmpliTube, you may want to use a combination. Most controllers communicate their MIDI with the computer software through USB, though a handful still use MIDI cables.

So what does this have to do with guitar?

Most of us have become accustomed to our usual, hardware-based world of amps and effects, where sounds change when we turn knobs or step on switches. Though the knobs and switches in AmpliTube respond easily to a mouse push or click, there is something satisfyingly familiar about adjusting the amount of fuzz by turning a knob and turning on a pedal by stepping on a footswitch. For fine adjustments, you will find that a MIDI controller knob is smoother than a mouse, letting you dial in the exact setting more easily. AmpliTube even lets you modify the curve of the adjustment to suit the situation.

Manual Controllers

The popularity of computer-based music has given rise to boom in MIDI controllers. These range in size from the Korg nanoKONTROL (see Figure 3.1) to units the size of small studio mixing boards—in fact, some mixing boards are in essence MIDI controllers. Manual-based controllers can have buttons, sliders, knobs, or all three—and more, such as a keyboard, percussion pads, or x/y pads.

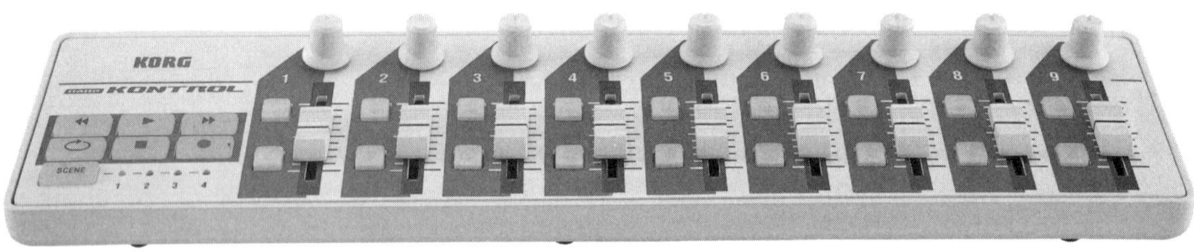

Figure 3.1 Don't let its 12.60" (W) × 3.23" (D) × 1.16" (H) size fool you; this is a powerful controller.

The size controller you need depends on how many knobs and switches you want to be able to access at once. Between its knobs and sliders, the tiny nanoKONTROL will allow you to manipulate up to 18 parameters (which doesn't count mapping one control to multiple parameters) and turn 18 effects on and off. It also allows you to change among four "scenes," each with different mapping. This means you can multiply all those eighteens by four, giving you control over 144 parameters, still without any doubling up.

As you'll see, even if you choose to use a foot controller, having a manual MIDI controller as well is not a bad idea. Here is an example of how you might set up the nanoKONTROL with AmpliTube.

Example 1

- In standalone mode, I set up an AmpliTube preset with an American Tube Clean 1 amp, fronted in the Stomp section by a Fuzz Age fuzz and an Echoman delay.

- Control-clicking on my Mac (right-clicking on a PC) over the amplifier's volume made a screen pop up with an option for Assign Midi.
- Holding the cursor over that screen revealed another screen with a Learn Volume option. (If I had clicked on the treble control, this option would be called Learn Treble.)
- Clicking on Learn Volume opened a screen labeled MIDI Learn that said, "Waiting for MIDI input." When I turned Knob 1 on a Korg nanoKONTROL, the screen disappeared, and the control lesson was learned. Then when I turned the nano's Knob 1, the Volume knob on AmpliTube's amp turned as well.
- Clicking on Stomp A in the routing area, I put a Fuzz Age in Slot 1 and an Echoman in Slot 6. Using the same process as with the amp's Volume knob, I set the top button in the nano's second channel to turn the fuzz on and off, while setting the knob in that channel to operate the Fuzz knob. I then programmed Channel 7 on the nano to control the Echoman: the top button for on/off, the knob for delay time, and the slider for feedback.
- Clicking on the MIDI button in AmpliTube's lower-right corner opened the MIDI Control panel (see Figure 3.2). There I could see all the assignments at a glance on the Preset page.

You will notice that while the delay, feedback, and fuzz controls show minimum settings of zero and maximums of 100 percent, the fuzz and delay bypass settings are reversed. This is a common MIDI issue: When the on/off switches were set with minimums of zero and maximums of 100 percent, the nano's switch lit up when the device was off rather than on. Switching the values brought them into synch.

The advantages of using a manual controller with AmpliTube are many:

- There is the aforementioned increase in the finesse of the adjustments. When you're setting delays manually rather than synched to a clock, it makes it easier to find the precise relation to the song's tempo.
- You will find that if you want to immediately switch a device on or off, it is much quicker to push a button than to move your mouse over it and then click.
- If you want to take advantage of the Echoman's Echoplex-style runaway feedback while simultaneously shifting the speed and pitch with the Delay knob—well, this is something you simply cannot do with a mouse and cursor.
- You can now adjust the volume of the amp without having to switch to the amp screen. One of the best things about a MIDI controller like this is that you can adjust parameters on all of the screens without having to leave the one you're on.

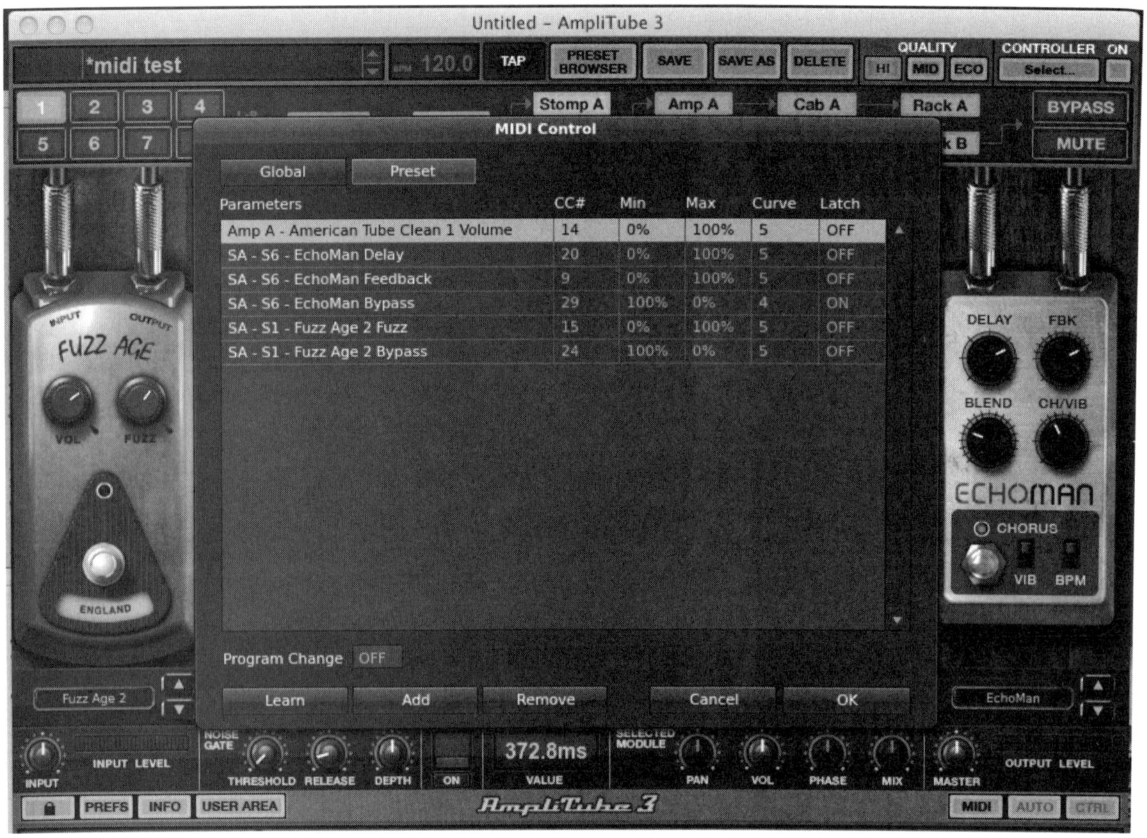

Figure 3.2 The MIDI Control panel shows you all of your MIDI assignments.

Imagine being able to adjust the mic placement while tweaking the amp tone without having to switch screens.

MIDI Controller Options

When searching for a MIDI controller, you will find that while some are called *foot controllers*, there is no category called "manual" controllers. MIDI controllers that are not foot controllers essentially fall into five categories.

- **DJ-oriented controllers.** Some of these are suitable. You may not use the crossfader (or find another use for it), but if there are enough knobs and switches, a DJ MIDI controller might work.
- **Launch pads.** These are geared toward Ableton Live users. They are for launching clips, and although their cute colored buttons can be used as switches, they are short on knobs and faders.
- **Beat-oriented controllers.** These focus on pads that you can tap for beats. Though these, too, can act as switches, the units tend to be short on knobs and faders as well.

- **Keyboard controllers.** Some, such as the Novation ReMOTE SL, sport dozens of buttons, knobs, and faders and can be just the thing, especially if you also play some keyboards in addition to guitar. Even if you don't, the keys can be set to work as on/off switches also.
- **MIDI controllers.** These may have some pads but no keyboard or crossfader. They are long on knobs and buttons. They are sometimes oriented toward controlling a software DAW mixer but can be ideal for controlling the myriad parameters of AmpliTube.

Some recommended controllers are:

- The Korg nanoKONTROL described and pictured earlier packs an enormous amount of control in a tiny, inexpensive package.
- Novation ZeRO SL MkII USB MIDI Controller (Figure 3.3) is a non-keyboard version of their Remote SL, and is likewise festooned with control options.

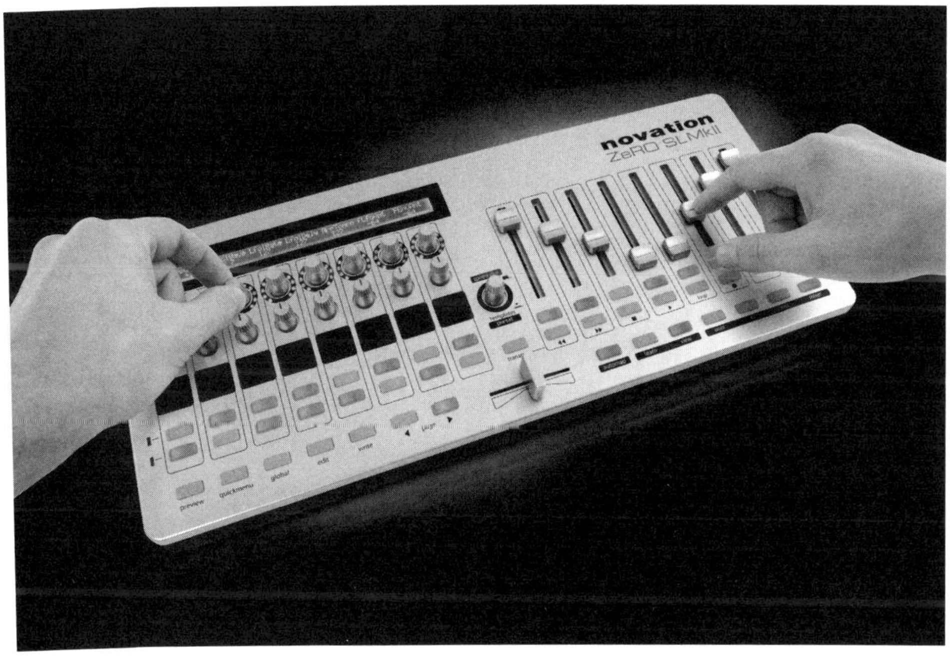

Figure 3.3 The Novation ZeRO SL MkII USB MIDI Controller makes a great addition to a foot controller.

Foot Controllers

If you never use wah-wah or whammy sounds, you can conceivably get by with just a manual controller. Still, odds are good that at some point you will want to use one of those effects. Jeff Beck is able to manipulate his Stratocaster's vibrato arm while doing swells with his guitar volume, but the rest of us mere mortals need a volume pedal to pull off that trick.

Express Yourself

At the very least, you'll want an expression pedal to handle wah, whammy, and volume effects. In addition to being an audio interface, the IK Multimedia StealthPedal (refer to Figure 2.5 in Chapter 2) is just such an expression pedal. It also has an expression pedal input so you can add a second pedal to control volume and wah separately.

Example 2

You can control two or more effects simultaneously with one pedal—and not just wah, whammy, and volume.

- Imagine being able to make the tremolo effect go faster as your wah-wah gets brighter. Check out Figure 3.4—you will see that the Opto Tremolo Rate and the Wah 46 Wah are both being controlled by an expression pedal sending out MIDI message CC#4. Both minimums are zero, and both maximums are at 100 percent. Thus, as the wah gets brighter, the tremolo gets faster.
- As you learned earlier, switching the minimum and maximum values causes a controller to produce the opposite effect. In this case you could use this feature to make the trem get slower as the wah-wah gets brighter. In Figure 3.5 the percentages are reversed; now the tremolo speed gets slower as the wah tone gets brighter.

Figure 3.4 The Tremolo Rate and the Wah are both controlled by MIDI message CC#4.

Figure 3.5 Here the percentages are reversed.

Perhaps a more common use of mapping the same pedal to more than one effect would be with delay.

Example 3

Here I have set it up so that with the pedal back, I get a nice slap delay for rhythm, and with the pedal forward, I get a longer delay for lead.

- In Figure 3.6 you can see that I have mapped three delay parameters to a single MIDI control: CC#7. (I have mapped this same control number to the Pro Drive, but we'll get to that shortly.)
- The amount of delay and feedback and the blend between the dry and wet signals are all controlled by the same sweep of my expression pedal. But I don't want them all to do the same thing, so I have carefully adjusted their minimum and maximum percentages to reflect the effect I want.
- The delay parameter's 15 percent minimum gives me a 96-millisecond slap delay, while its 100 percent maximum affords me the full, almost 500 milliseconds.
- For slapback, a single repeat is usually enough. Setting the minimum for 0 percent gives me one repeat. For longer delays, I prefer a few more repeats. Trial and error reveals that 43 percent gives me about four or five—perfect.

Figure 3.6 Three delay parameters mapped to a single MIDI control/settings heel down.

- When using a slap delay, I want to be able to hear it a little more in the foreground, so 27 percent works for the blend. Reducing it to 15 percent at the maximum setting puts the longer delays in the background, out of the way of my lead work.
- As I am using the pedal to change from a rhythm to a lead tone, I want the Pro Drive to have more distortion for my leads than for my rhythm work. Mapping the Pro Drive's distortion control to the same CC#7, I can set it to increase when I push the expression pedal toe down. Figure 3.6 shows the knob settings with heel down, while Figure 3.7 reflects the knob changes when I push the toe down.

Switch Master

You may already be familiar with hardware multi-effects with presets that switch a number of effects on simultaneously, while shutting others off. With AmpliTube and a foot controller, you can either step through presets that offer the desired effects or set up Stomp A and Stomp B as one big pedalboard (or a combination of both), assigning one switch to turn on a single pedal or multiple pedals at once.

Example 4

- In Figure 3.8, note on the MIDI Control screen (now moved toward the bottom so you can see the pedals) that the minimum/maximum percentages are one way on the delay and fuzz and reversed on the phaser and tremolo.

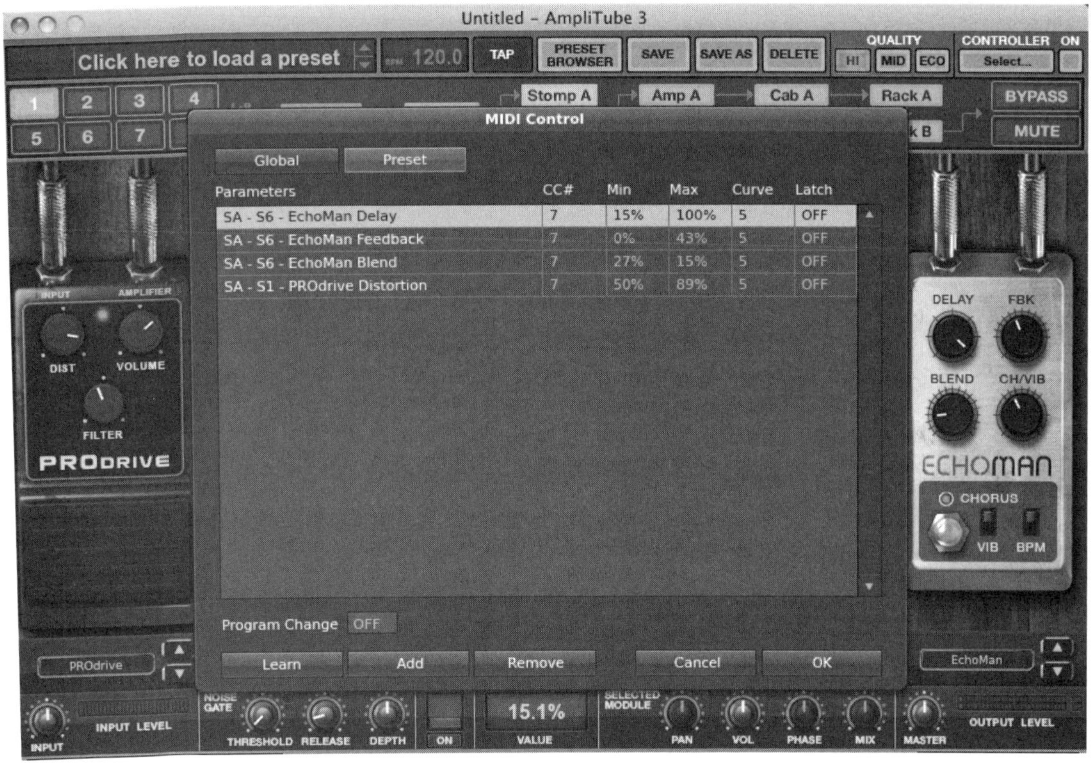

Figure 3.7 Stomp settings when the expression pedal is toe down.

Figure 3.8 All four bypasses are controlled by a footswitch sending CC#24.

- All four bypasses are controlled by a footswitch sending CC#24. This means that when I step on that single footswitch, the fuzz and delay will go on, while the phaser and tremolo will be shut off.

Foot Controller Options

There are a number of foot controllers out there, but these are a good place to start.

- The IK Multimedia StealthPedal's audio interface attributes were covered in Chapter 2. It is also the perfect starting place for controlling AmpliTube's effects and functions. The built-in expression pedal and switch can be associated with any parameter of "Powered by AmpliTube" software or any other amp and FX modeling software with MIDI control capabilities. The StealthPedal controls are expandable with input for an additional expression pedal and two extra switches, allowing you to build a mini-pedalboard rig with preset and single effects control capabilities for live/recording usage. It offers a Volume knob and three LED indicators that operate as a status display, tuner, or level indicator when used with any "Powered by AmpliTube" software.
- The Behringer FCB1010 (see Figure 3.9) has been an extremely popular floorboard MIDI controller for years among MIDI-minded musicians. It is inexpensive and rugged. It offers 12 mechanically silent switches (silent is important in quiet musical situations), with inputs for two more, and two expression pedals.
- The Line 6 FBV Shortboard MkII (see Figure 3.10) integrates nicely with AmpliTube. It has one onboard expression pedal and an input for a second.
- The IK Multimedia StealthBoard (see Figure 3.11) is of course the easiest to match up with AmpliTube. It integrates fully with the StealthPedal, instantly adding 10 assignable switches and a second expression pedal. It controls "Powered by AmpliTube" products with no setup—simply plug your computer into the USB port of the StealthBoard and immediately access thousands of "Powered by AmpliTube" presets and parameters. It also has four additional inputs that can be used for either more switches or more expression pedals. The StealthBoard features MIDI in/out connections for any other MIDI-controllable gear, including MIDI-controllable amplifiers, making it great for integrating AmpliTube with real amps either live or in the studio.

Control Considerations

What sort of controller you need will depend on how you plan to use AmpliTube. If you plan to use it just for recording, a StealthPedal and a manual controller should give you quick access to everything you need. MIDI mapping can be set for global use, which means that the expression pedal(s) will always control the same effects, and the switches turn on and off the same stomp positions, regardless of the chosen preset—or their destination (the parameters that they control) can be set to change

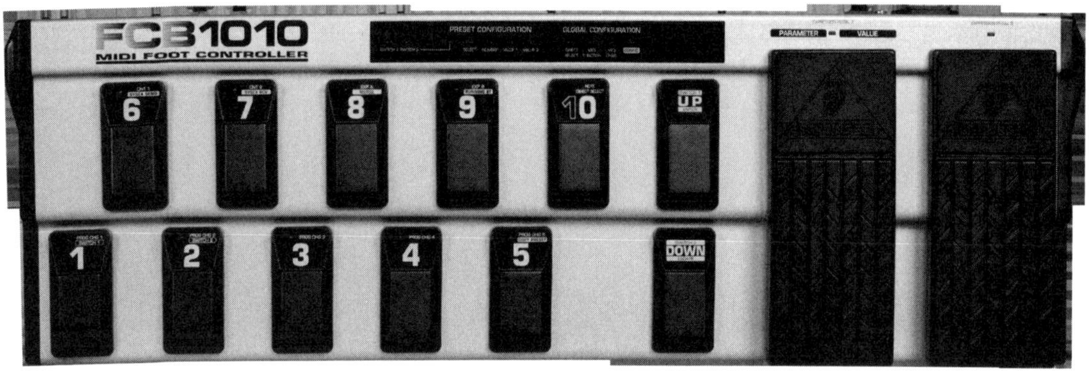

Figure 3.9 The Behringer FCB1010 is one of the more popular floor-base MIDI controllers.

Figure 3.10 Originally designed for Line 6 amps and PODs, the FBV works well with AmpliTube.

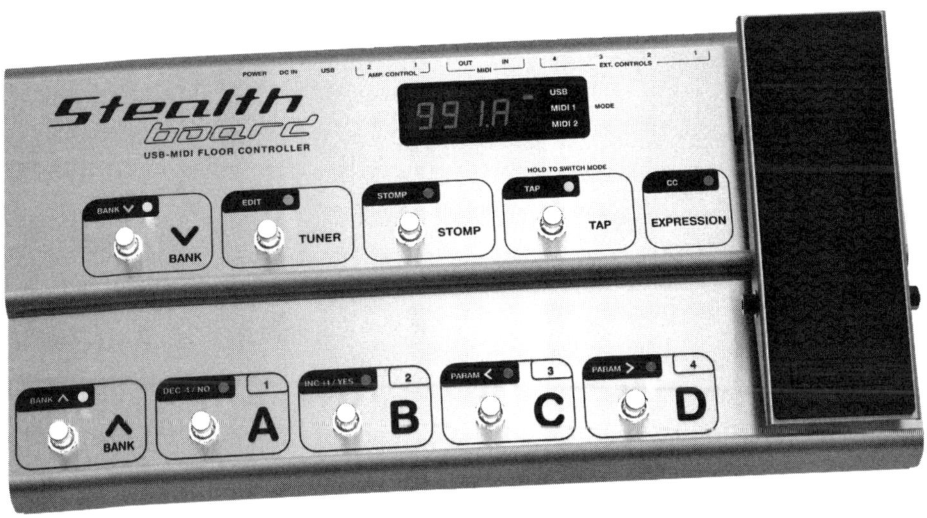

Figure 3.11 The IK Multimedia StealthBoard integrates immediately with AmpliTube.

with each new preset. You'll find that the sonic options multiply significantly once you take MIDI control of AmpliTube, whether with your hands or feet.

Summary Questions

You'll find the answers to the Summary Questions in the Appendix at the end of this book.

1. What does MIDI stand for?
2. What does CC stand for?
3. You can map a single controller to:
   a. One parameter
   b. Two parameters
   c. Any number of parameters
4. Reversing the minimum and maximum percentages of a setting on the MIDI Control page has what effect?
5. How do you make one effect turn off and another turn on using one switch?
6. Give two reasons for using a manual controller with AmpliTube.
7. Which effects require an expression pedal?
8. On which effects can you use an expression pedal?
9. Why is a beat-oriented pad controller less than ideal for AmpliTube?

4 Monitors

By this point I'm sure you're itching to get into the gear, but bear with me for one more short chapter. After you've decided how you will get your sound into your computer and determined a method to control AmpliTube's software, you need to figure out how you intend to listen to the sounds produced—what are you going to use to monitor your audio?

As with interfaces and controllers, the answer will be dictated by circumstances and pocketbook, but there are principles that will help ensure that the way you hear your tone in your head is the same way that you and others ultimately hear it.

Monitoring audio is a big subject, practically worthy of a book of its own. In this chapter, I will try to take a briefer, guitar-centric approach.

Speakers or Headphones?

It seems that these days people do most of their music listening on headphones, whether the music is on a laptop, an iPhone, an iPod, or another personal music player. These headphones range from the free earbuds provided with the players to pairs that compare favorably to high-end studio monitor speakers. An extremely large portion of music consumption is also done in cars, commuting to and from school and/or work—and yes, people still listen to their tunes through speakers in their homes, whether those speakers are audiophile systems or cheapo computer models. Therefore, it is important that you ultimately check your tone through both headphones and speakers—the more different kinds, the merrier.

Headphone Advantages

- Headphones are ideal for checking stereo panning, hearing unwanted noise, and tending to sonic detail.
- If you work around distractions, such as family or roommates, headphones help shut them out.
- Headphones save said family or roommates from having to hear you play the same lick over and over again.

- Headphones allow you to work late at night without having to own a private house or a soundproof studio.
- Headphones eliminate anomalies, such as boominess, and excessive liveness or deadness of any room.
- Headphones are comparatively cheaper than speakers.

Headphone Disadvantages

- Headphones do not let you know how a room will affect the sound. When heard through speakers (or in live performance), music bounces off the walls of the room, causing phase cancellation that masks certain frequencies. This doesn't happen with headphones, so your EQ choices will be different.
- Headphone panning will sound dissimilar when played though speakers. Panning tends to sound more exaggerated through headphones.

Speaker Advantages

- Speakers give a good sense of what the music sounds like in a room.
- Speakers usually provide better, more accurate bass information and EQ information in general.
- Speakers help you "feel" the sound more than headphones do.

Speaker Disadvantages

- Speakers can be annoying to neighbors, family, and friends.
- Subtle noises, such as clicks, pops, string squeaks, and amp hiss, are not as easily detected when monitoring through speakers.
- Speakers are generally more expensive than headphones.

Choosing Your Poison

As you have learned, both speakers and headphones have good and bad points. Depending on your needs and budget, you will tend to work more through one than the other during your day-to-day workflow, resorting to the alternate for spot checks. When buying either or both, you should apply certain principles.

- Listen carefully before you make your purchase. This seems obvious, but with the prevalence of searching for the best deal online, you may forget that you have to listen to the product before you buy. Many stores offer the opportunity to compare and contrast—take the time to do so.

- Listen to them playing a musical source that features a wide frequency spectrum. You will want to make sure that the speakers or headphones are even throughout, without emphasizing bass, mid, or treble.
- Listen to a high-quality (CD, not MP3) musical source that you know extremely well and have listened to under many circumstances. This will help you judge the accuracy of the monitors.
- Listen loud, listen soft, and listen at levels in between. The best monitors will produce the same frequency balance at any volume. Listening loud will ensure that there is no distortion (the bad kind) happening in the drivers.
- As a guitarist, you should, of course, choose source material that features the kind of guitar tone you prefer. Make sure that it is reproduced accurately.

The Bottom Line

Many terrific sources will help you set up monitor speakers to the best advantage in your home studio. The point that I would like to leave you with here is that familiarity with your main monitoring system, whether it is headphones or speakers, is ultimately the most important factor in producing great tones that transfer well to other systems. I recommend that you do as much of your music listening as possible through the computer on which you plan to record and through the monitors you use to record. By getting used to what great tone sounds like through that system, you will find yourself working to reproduce that same tone.

Whether your favorite guitarist is Jeff Beck, Dimebag Darrell, Brent Mason, The Edge, Bill Frisell, any other axe-wielder out there, or all of the above, if you work toward making sure your guitar sounds are similar to theirs, you are bound to be in the ballpark. Their recordings are mixed and mastered to sound great on a wide range of speakers. If your tone equals theirs on your system, there is a good chance that it will translate to others.

Another consideration is that you like the sound of the primary monitoring system you choose. It doesn't matter whether a magazine, a world-famous guitarist, or a producer recommends it; you are the one who is going to be listening to it for many, many hours at a time. If you use headphones, make sure they feel good on your head and ears.

That said, it is important that you use speakers and/or headphones specifically designed for recording if at all possible—and the best quality that you can afford. (You should check your tone on earbuds, but working on them—no.) Companies such as Sony, Sennheiser, AKG, and Ultrasone make headphones that will minimally color the sound, while studio monitors are available for Genelec, Tannoy, Yamaha, M-Audio, KRK, and others in a wide range of sizes and price points.

Be sure the sound coming out of a pair of headphones or speakers is pleasant and inspiring, and you are halfway there. Become intimately acquainted with your monitoring system's sound, and you are all the way home.

Summary Questions

You can find the answers to the Summary Questions in the Appendix at the end of this book.

1. What are two advantages to monitoring through headphones?
2. What are two disadvantages to monitoring through headphones?
3. What are two advantages to monitoring through speakers?
4. What are two disadvantages to monitoring through speakers?
5. What are two things you should do when purchasing your monitoring system?
6. What is the most important thing to remember about a monitoring system?

5 The Amp

At last, we get to the place where your great guitar tone will be fully shaped—the cool gear in AmpliTube. There is every chance that the future will find you recording and perhaps playing live through this software, but even if you continue to use hardware amps and effects for recording and onstage, working with AmpliTube will help you learn which gear you need to achieve the awesome tone you seek.

As a long-time Fender amp man, I had much less experience with Vox and Marshall products before I began working with modeling software. AmpliTube let me experiment extensively with those sounds, as well as Orange, THD, and Mesa Boogie tones, among others. (I eventually added an Orange amp to my arsenal.) As you will learn, each amp enjoys its own tonal characteristics and responds differently to various effects.

All the amps available in AmpliTube Jimi Hendrix, Metal, and AmpliTube 2 have been remodeled and added to AmpliTube 3.

Context Get ready: I'm going to impart the most important secret to getting great guitar tone:

Great guitar tone exists only in context.

I will be emphasizing this over and over throughout our journey. An appropriate tone when playing solo guitar is not necessarily the best tone for playing with others. Playing in your room, by yourself, you will revel in gobs of low end emanating from your amp, while luxurious reverbs and delays fill the room with glorious ambience. By all means, enjoy it in that context, but be warned: As soon as you start playing with others, whether in live performance or on record, that tone may have to change. All of that bottom will merely compete with the bass player's sound, rendering you both indistinct at best and inaudible at worst. That lush ambience that makes your leads reverberate around your basement will sink them way back in the band mix, causing your friends to come up to you after the gig and go, "The band was great, but I wish I could've heard your solos better." Often you'll find that the proper sound for a song may seem thin and dry on its own, while fitting in the mix like a gear in a Swiss watch.

The best way to approach the amplifiers in AmpliTube is to go through the models one at a time. With each one, I will tell you a little bit about the historical significance of the original amp represented by that model. Then I'll discuss the nature of its tone, as well as how to get the most out of it. Along the way, I'll impart general tone tips as well as specific ways to deal with each model's quirks.

By all means, go through all the presets provided for each amp model. You'll find much gold there, and you may find just the sound you want right out of the box. But for the purposes of this book, I will often have you start from a neutral setting and demonstrate how to build up a particular sound.

American versus British Throughout this chapter, I'll be referring to amps as having a more *British* or a more *American* tone or sometimes a combination of both. In the amp world, American tone has come to represent a certain smooth warmth, resulting from a big low end and an emphasis on the low mids, whereas British tone conjures up an edgier sound, with a hint of metallic clang, produced by strong upper mids and a sharper high end, as well as a tighter bottom.

One last thing before we start: It's essential that you keep an eye on AmpliTube's input and output meters. If the tone ever starts to sound distorted in a bad way (and not in a *good* "bad" way), check the meters. More than likely you'll find that the input, the output, or both are hanging out too long in the red zone.

Now let's get into gear.

American Amps

This section will cover the models based on amplifiers manufactured by classic American-based companies.

American Clean MKIII

Randall Smith, creator of the Mesa Boogie amp company, was an amp repairman known for his hot-rodding of Fender Princetons. He would take this 12-watt amp with a 10″ speaker and goose it up to 50 watts with a 12″ speaker. The warm, clean sound and singing lead tones of these amps impressed Carlos Santana, as well as a generation of Los Angeles session players, such as Larry Carlton, Lee Ritenour, and Robben Ford.

When he launched Boogie, the amps added switchable channels, 5-band EQ, effects loops, and a wealth of knobs to control the lead and clean channels, making Boogies a tone-tweaker's dream. Modern players, such as John Petrucci of Dream Theater and Metallica's James Hetfield, have also embraced Boogies.

AmpliTube has separated the MKIII's two channels into two models. The American Clean MKIII offers the classic warm, clean sound for which Boogie is known, as well as the tone-shaping abilities of the original.

Exercise 1

Here we'll build up a cool, clean sound, perfect for funky rhythm work and chiming arpeggios.

Let's start with the model set like Figure 5.1. Take the MKIII Clean 1 setting and shut off the Bright and Deep buttons. If, like me, you're playing a Stratocaster with a maple neck through it, this might sound a little anemic. If you were playing a Les Paul through this same setting, it might sound just fine—context! For the purposes of this exercise, let's assume you're playing a single-coil guitar, such as a Strat or a Tele.

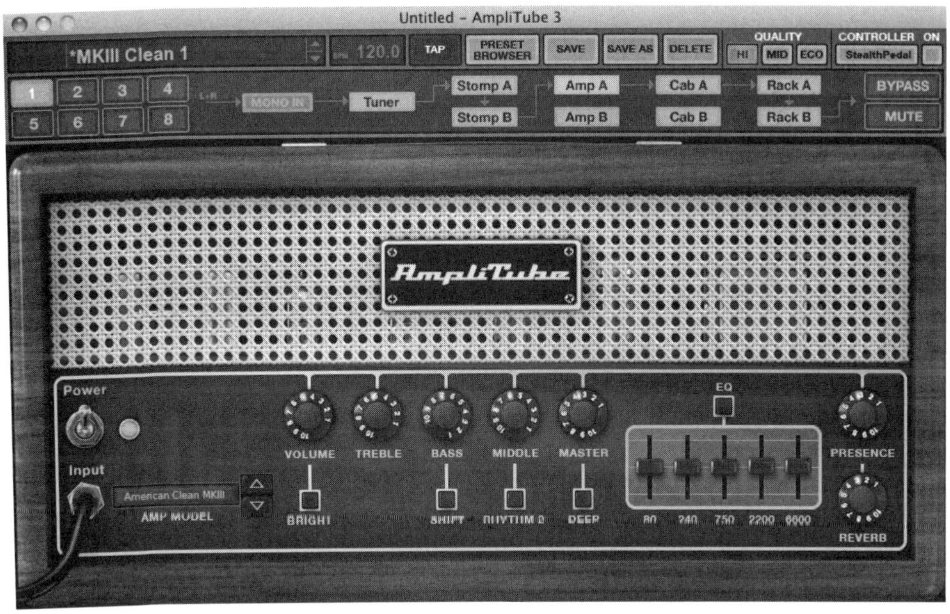

Figure 5.1 Start by building clean tone from a clean slate—all boosts and graphic EQ off.

1. Before you touch any of the tone controls or buttons, turn up the Volume knob from 4 to 6. That is immediately better—not just louder, but fuller. As the preamp moves toward distortion, it acts almost like a tube compressor fattening the sound and adding sustain, but still shy of any audible grit.
2. If it still sounds a little dull, turn on the Bright switch and turn up the Presence. We now like the amount of high end, but the treble has a bit more edge than we want. Refrain from twiddling with any more tone controls or turning on the equalizer; instead, go to the cabinet section and move Mic 1 away from the center of the speaker toward the edge. If you feel that you've lost some of the presence, you may have gone too far. Split the difference, and there it is: a warm, present, clean tone.

3. It sounds great on its own, but there is a bit of low-end woof that could prove problematic in a live or recorded mix, when combined with a bass player's frequencies. Roll down the bass from 7 to 5—now it's just right.

Of course, that's just right for my guitar and my touch. Your guitar may sound darker or brighter, the way you hit the strings may be harder or softer, and your strings or pick may be heavier or lighter gauged—all of this will affect the way you modify the controls and mics.

American Lead MKIII

This model is based on the lead channel of the same amplifier. AmpliTube doesn't offer one head that lets you switch channels like on the original combo. This is better, as it allows you many more sonic options. One disadvantage of the original amp was that the same tone settings had to work for both lead and rhythm sounds. Here you can have entirely different tone configurations for each. If you wish to switch between American Lead MKIII Lead and Clean heads, you can create a Clean MKIII in one preset and a Lead MKIII in another.

Figure 5.2 shows an MKIII Lead setting that sounds good with my guitar on its own—but what about when playing in a band?

Figure 5.2 A lead setup that sounds good on its own but maybe not in a mix. Note that the EQ is off.

Among the multitude of uses for AmpliTube's 4-track audio player/recorder with Speed Trainer is the opportunity to load in a backing track and see how the preset sounds in the context of a mix. Unless you plan to play that preset in a solo guitar performance, you'll want to check how it sounds with bass, drums, and keyboards.

Exercise 2

This exercise will let you see the difference between a sound on its own and a sound in the mix.

1. Set up the American Lead MKIII in the configuration pictures in Figure 5.2. Set the EQ as pictured but be sure not to turn it on.
2. Now add a full stereo backing track to the audio player on Track 2, as pictured in Figure 5.3. Pick one that you feel might call for a sound like this. If you don't have one, there are many available online.

Figure 5.3 Here's a backing track loaded into Track 2. *Do not* add AmpliTube to the track.

3. Play along with the track. Hear how the low end and low mids conflict with the other instruments.
4. Now turn on the EQ and play some more. The guitar now blends better into the track. You'll also find that you can lower its volume in relation to the track and

still hear it. You won't miss the lows that have been rolled off, as there is plenty of bottom coming from the bass and the bass drum.

5. If your lead still sounds separate from the backing tracks, you may want to add some reverb or a little delay.
6. Record a track playing through each version (with and without EQ)—one to Track 1 and one to Track 3. Listen to each, muting the other track. You will find it is easier to concentrate on hearing the difference when you're not playing.

American Tube Clean 1

This model is based on the "Blackface" Fender Super Reverb, an amp that Fender produced between 1963 and 1968. The vintage versions employed two 6L6 tubes to provide 45 watts. Stratocaster-wielding blues players have long favored them, as the way they drive their 4×10″ speakers wired to 2 ohms creates just the right breakup for that Texas-blues tone. AmpliTube 3 is one of the rare places that you find this amp modeled—even AmpliTube Fender doesn't offer it. To achieve this classic sound, make sure you start with one of the American Tube Clean 1 presets that employs a 50-watt 6L6 power amp.

Some presets pair the preamp with a 100-watt power amp; this sound is closer to the post-1968, 80-watt "Silverface" Super Reverb.

Preamp and Power Amp This a good place to discuss the difference between the preamp and power amp stages of an amplifier. There are plenty of sources that can explain the technical workings of each and how they interact electronically. Here I will talk about the tonal effect of each.

The Preamp The preamp stage of a head or combo is where much of the amp's character lies. The way the circuit is designed and the way the tone controls interact determine a large portion of the way the amp sounds. Two amps with the same preamp circuitry but a different number or type of power tubes will definitely sound different, but two amps with the same power section and different preamp layouts will differ more. Early Marshall amps used very similar power stages to Fenders, but their preamp and tone circuitry made all the difference.

The Power Amp The louder you play, the more effect the power amp has on the tone. At lower volumes, the power tubes are mostly conveying the character of the preamp section; as you get louder and start to distort the power tubes, the way those tubes break up will add more and more to the tonal character of the amplifier.

Exercise 3

Try this to hear the difference in sound and response that two extra 6L6 tubes make.

1. Set up an American Tube Clean 1 Bluesy Combo to look like Figure 5.4.

Figure 5.4 This configuration should produce a vintage blues sound.

2. First lower the Reverb a bit. Just like the originals, the reverb on AmpliTube's Fender models comes on very fast, and by the time you pass 2, it's more surf than blues.
3. Reduce the Bass to about 2 1/2. This is not just to fit better in the mix. Also like the originals, these vintage Fender combos get very loose and flabby if the bass is up much more than this.
4. Raise the Middle a couple of notches. Midrange tends to punch an amp into distortion sooner than high end. This is one reason why P-90 pickups break up an amp faster than Strat or Tele single-coils.
5. Dime out the volume; we're going for as much drive as this amp can produce.

   This should produce a slight breakup when you dig in hard with single-coils and a little more breakup with humbuckers. Backing off the volume on your guitar will give you a clean sound.

6. Now, without changing any other settings, change the power amp to 100-watt 6L6. You'll hear and feel a difference. The amp won't break up quite as much when playing chords, and the feel will exhibit somewhat less "sag." The feel and the low end will be tighter than with a 50-watt power amp. The soft sound produced by a "Blackface" Super Reverb is desirable for some types of old-school blues, but if it's not what you want, feel free to use the 100-watt option.
7. Finally, go through all the other power-amp options without changing anything else. You will hear a wealth of subtle (and not so subtle) sonic and feel changes with each new type and configuration of tubes (see Figure 5.5).

Figure 5.5 Each power-amp configuration offers a slightly different sound and feel.

American Tube Clean 2

Tube Clean 2 is based on the 1965 "Blackface" Fender Deluxe Reverb. This 1×12" combo's two 6V6 tubes put out about 22 watts of power, making it perfect to crank up in small clubs or for keeping the stage volume down while miking it in larger venues. A favorite of pop, country, and blues players who want that pristine Fender chime at lower volumes, it is only roughly approximated in AmpliTube 3.

The American Clean 2 preamp and EQ sections capture much of the original's chime and bite, but there is no 6V6 power-amp option available, and the 50-watt 6L6 offered produces a much more open sound with higher headroom than the original. It is a wonderful sound—just not quite a Deluxe Reverb. If you're a stickler for this particular tone, you should go for the AmpliTube Fender package, where it's nailed perfectly.

That said, the American Clean 2 option provides a distinctly different character than Clean 1, and you should try it with all of the available power-amp options. You will likely find some combinations you can't live without.

American Tube Vintage

This amp is based on the "Tweed" Fender Bassman 4×10 combo from 1959. Originally designed to be a bass amp for the revolutionary Fender electric bass, within a few years this 50-watt amp was coveted by guitarists for its gritty breakup when pushed. The specs appear very similar to the Super Reverb—two 6L6s, 50 watts, 4×10" speakers—but the sound is quite different.

Exercise 4

Hearing is believing, and it's always a good idea to get used to the sound of each amp, so let's set up an instance of AmpliTube using Routing Preset 2. This will allow us to compare and contrast the sound of these two models.

1. First, install an American Clean 1 in Route A, with no effects, and a 50-watt 6L6 power-amp stage. Set the controls to Bass at 2; Middle, Treble, and Presence at 7; and the Master volume at 8. Turn off the reverb. (You should always adjust your tone with the reverb off first, as it can make you settle for a lesser tone than you otherwise might.) Rule: If it sounds good without reverb, it really sounds good.
2. Next, place an American Tube Vintage amp in Route B, making sure that the settings are identical, down to cabinets and mic choice and placement. (You can do this by right-clicking or Control-clicking on the cabinet in A to cut and paste the same configuration into B.)
3. Turn off the amp in Route B (Bassman) and play through the one in Route A (Super Reverb).
4. Now turn off amp A and turn on B.
5. Note how the American Tube Vintage is darker and more broken up. This darkness is a hallmark of the Tweed-era Fender amps.

American Tube Vintage B

Fender followed the coveted 4×10" Bassman combo with a separate head and cabinet version a few years later. The closed-back cabinet housed a more bass-appropriate 2×15" configuration, but this didn't keep it from ultimately finding more use among guitarists than bassists (though, if not driven to distortion, it makes a great bass amp).

The version in AmpliTube 3 comes from the now-included AmpliTube Jimi Hendrix series. As with the AmpliTube 3's Fender Deluxe model, it varies significantly from the original hardware but maintains the spirit of the tone. Though never offered on the vintage version, this Bassman sports a Gain control to help you dial in your desired level of distortion at any volume. Though Fender later offered 100-watt versions of the Bassman head, the '60s version was a twin 6L6 50-watt edition. You can easily emulate this by switching the power section to the appropriate configuration, though you might enjoy the extra headroom of 100 watts. There was no reverb on the original, but spring reverb is always welcome on a Fender amp.

This is not just a head version of the 4×10" combo with different speakers. The Bassman heads offer a darker, smoother sound than the combos—one that AmpliTube has amply captured. You can and should hear this for yourself by trying an A/B test.

1. Using Routing Preset 2, place an American Tube Vintage in signal path A and an American Tube Vintage B in signal path B, shutting off both Spring Reverbs.
2. Change the speaker configuration for the American Tube Vintage B to 4×10" Open Vintage and the power amp to 50-watt 6L6.
3. Make sure the mic setup for both cabinets and the amp volume and tone settings are the same. (Having the American Tube Vintage B Gain at noon should let you approximately match the master volume sound to American Tube Vintage.)

You should be able to hear a subtle yet distinct difference in the character of each amp.

American Tube Vintage D

Another import from the Hendrix collection, the American Tube Vintage D models the Fender Dual Showman. The Fender Showman is essentially a Fender Twin in piggyback head form. The Showman came with a 1×15" cabinet, while a 2×15" bottom accompanied the Dual Showman.

As the bands of the '60s got louder, before Marshall and Vox established a foothold, Showmans and Twin Reverbs were everywhere. They were valued then and now for their warm, clean tone. Funk, country, and blues players enjoyed the extra headroom that allowed them to play loudly without distorting. Still, if pushed hard enough by humbuckers, the early "Blackface" models would break up and sustain, creating a smooth distortion for players such as Santana and Peter Green. The premier boutique amp of all time, the Dumble, is largely based on this circuitry.

The American Tube Vintage D head appears loosely based on a Dual Showman Reverb model with a Presence knob instead of a Bright switch for more control of those upper frequencies, and a Gain knob that lets you dial in the desired preamp "push" at any volume. The 2×15" configuration of the cabinet offers not just enhanced low end, but also a more even frequency response over the spectrum from high to low. You can really push the high end on this model without encountering any harshness. If you want a little more sparkle without bite, try switching the EQ section from American Tube Clean 1 to American Tube Clean 2.

American Tube Vintage T

Also used by Hendrix in his early days was the Fender Twin Reverb (see Figure 5.6), modeled in the American Tube Vintage T. Once again, AmpliTube has added a

Figure 5.6 The Fender Twin: one of the favorite clean amps of all time.

convenient Gain control and Presence instead of a Bright switch. You may also notice that both this model and the Vintage T lack the tremolo available in the originals. (The AmpliTube Fender '65 Twin Reverb restores both the Bright switch and the tremolo.) Fear not, the Opto Tremolo in the Stomp section will supply all of the wiggle that you need.

- The default American Tube Vintage T opts for an American Tube Clean 2 EQ section. If you find the sound too edgy when combined with the 2×12 Open TJ120 cabinet's JBL speakers, I would start by switching the EQ section to American Tube Clean 1.
- Alternatively, you could try the 2×12"Open SL Silvertone speakers or the 2×12" Open Vintage, with its Vox AC30 '60s Bluebacks. Or, you could change both the EQ section and the cabinet—such are the joys of AmpliTube.

Whichever 2×12" configuration you choose, you'll notice that the Twin sounds more focused in the midrange than the Dual Showman with its 2×15" cabinet. Midrange is where rock guitar lives, making the Twin sound perfect for use with all manner of overdrives and fuzzes.

THD BiValve

This is a model of the THD BiValve head, a popular boutique amplifier that uses a less common single-ended class-A design for its power stage. (See the upcoming "Tube Talk" sidebar.) Once you get past the power stage, basic tone controls, and graphics (including a cool effect that makes the visible tubes glow brighter as you turn up the volume), the AmpliTube version of this head differs quite a bit from the original. Just as the Fender emulations substitute a Presence knob for the on/off original's Bright switch, here the AmpliTube THD replaces a High Cut switch with a Cut knob that allows more control over that function.

THD is the manufacturer of the Hot Plate, an attenuator that allows you to crank your amp's power stage and then takes the volume down between it and the speaker(s). The actual THD head comes with a built-in Hot Plate, which is unnecessary here, as AmpliTube provides the control over the master output.

Missing is the Attitude knob, a sort of presence control, and the More and Less inputs, offering different gain amounts. This model sounds as if IK has opted for the Less effect —though cranking the volume will yield aggressively crunchy chords and some solo sustain. For a simulation of the More input, I recommend using a Diode Drive from the Stomp section, with the Dist knob low and the Level high to push the BiValve into all its gritty glory.

Aggression is the name of the game with this amp, creating a breakup that is guaranteed to cut through the densest mix. You will find that adding the Single Ended Cl A power stage to any other amp model will significantly up the "cut" ante.

Tube Talk Because discussion of the THD amp alludes to things such as *Class A* and *single-ended design*, a brief, relatively nontechnical description of these concepts seems in order.

Two of the main types of output stages used by guitar amplifiers are single-ended and push-pull. This refers to the way that the amplifier's output stage circuit components connect and influences the type of transformer used.

Most guitar amps are push-pull and can be biased Class A or Class AB. Most designs are biased Class AB for efficiency and power output with minimal crossover distortion, though there is some debate as to whether the tone is as good as with Class A amps. Push-pull amps require a minimum of two tubes to operate.

The advantage of the push-pull circuit is that it allows smaller, cheaper transformers, while a disadvantage is that even-order harmonics and distortion tend to be canceled out, and more odd harmonic distortion is generated. Since even-order harmonics are found more pleasing by most guitarists, those that use the amp as their primary distortion source may prefer a single-ended design.

A single-ended guitar amp output stage is usually biased Class A, maximizing the output power before distortion—the Fender Champ uses a single-ended output stage. This design tends to be inefficient, producing much lower power levels than push-pull output stages, and it needs a physically larger and more costly output transformer. Single-ended circuits do not reject even-order harmonics—they actually emphasize even-order harmonics. This produces a sound that is highly prized by some players.

Tube Vintage Combo

This model mimics a late '50s Supro combo. Supro was one of many brand names manufactured by the Valco company between the 1940s and the 1960s. Jimmy Page is said to have recorded much of the first Led Zeppelin record through a Supro, and if you plug in a Telecaster, fire this baby up, and launch into "Good Times, Bad Times," it's hard to doubt that he did. Another single-ended Class A design, the Tube Vintage Combo plays especially well with overdrives and fuzz pedals, particularly the Fuzz-One (see Figure 5.7).

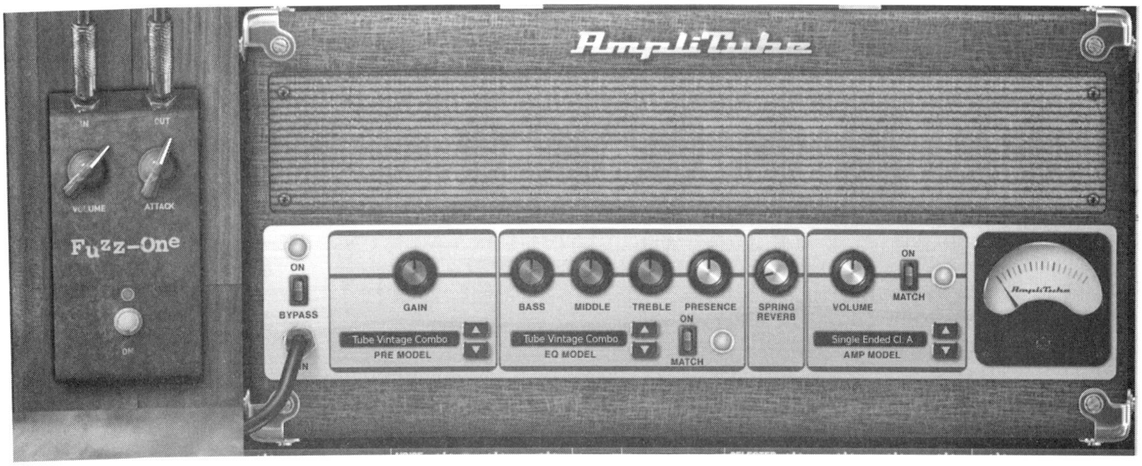

Figure 5.7 The Fuzz-One pedal pairs up nicely with the Tube Vintage Combo.

Modern Tube Lead

The rise of modern metal allowed American amplifier manufacturers to break the monopoly on hard-rock amps, solidified by the British in the '60s. Marshall, Vox, Hiwatt, and Orange weren't making amps with enough gain and aggression to satisfy the young players who were into speed, death, and various other genres of the new or nü-metal. Mesa Boogie leapt into the gap with its Rectifier series. Its 1989 Dual Rectifier first found favor with the grunge crew but truly came into its own when adopted by bands such as Korn, Limp Bizkit, and Linkin Park.

The Dual Rectifier allowed users to choose between a tube and a solid-state rectifier. The Modern Tube Lead emulation does not offer this choice and employs the 150W 6L6MT power stage from the Triple Rectifier rather than the original 100-watt power section of the hardware version. You can't argue with the results, however: The Modern Tube Lead captures the big bottom and more aggressive sound that made the Dual Rectifier the first American amp to catch on with the metal crowd. In AmpliTube the main difference between this model and the Metal Lead T or Triple Rectifier is a brighter EQ Model in the latter.

Metal Clean T

In the spirit of "more is more," Boogie followed up the Dual Rectifier with the Triple Rectifier, adding two 6L6s, 50 watts, and more channels. The Metal Clean T is based on the Triple Rectifier's clean channel. As with the MKIII, AmpliTube has broken this multichannel amp into its component parts.

Rocking 150 watts courtesy of six 6L6 tubes, the clean channel of this amp offers gobs of headroom.

- Though this amp is perfect for chorused metal arpeggios, don't be afraid to use its warm, undistorted sound for pop, funk, blues, or even country if it suits.
- Using 1×12", 2×12", or 4×10" cabinets instead of the metal-approved 4×12" style will help this versatile head adapt to many other styles.
- Consider changing the output tubes as well—in other words, dropping them to 100 or 50 watts for a little more give.
- Before you move on to the lead version (Metal Lead T), try putting one of the Metal Distortion pedals in front of the clean channel. You will find that it sounds even bigger than when placed in front of the lead channel.

Metal Lead T

Mesa Boogie likes to load its amps with options: knobs and switches that provide a plethora of different sounds. The hardware version of the Triple Rectifier sports 25 knobs and 5 switches—and that doesn't count the knobs that double as push-pull switches. The AmpliTube version has stripped it down to 7 knobs and 3 switches but still offers a wide range of tones with the now-classic Rectifier character. There are no switches for the Vintage/Raw/Modern options of the original, but those types of sounds are easily enough achieved.

- For raw and modern high-gain sounds, stick to the 150W 6L6MT power stage. Here both medium and high gain settings will yield an edgy, modern sound.
- The original Triple Rectifier allows the use of 6L6s or EL34s. Switching the AmpliTube model to a 100W EL34 configuration will instantly produce a more vintage sound.
- In general, more midrange equals vintage, whereas scooped midrange equals modern. The Metal Lead T's tone controls are ideally voiced to help mold a wide variety of head-banger sounds.
- If you want more gain but don't want to lose the character of the amps, consider adding a Diode Overdrive in the Stomp section.

- The Metal Lead T is very dynamic. If you keep the gain below 3 o'clock, it will respond to your guitar volume and playing touch, allowing you to adjust the distortion level by backing off your instrument level or adjusting how hard you play.

Metal Lead V

In 1991 the Peavey 5150 amplifier was designed by Peavey in conjunction with Eddie Van Halen, one of its endorsers at the time. With 100 watts of pure metal power, it was soon adopted by other artists seeking an amp that offered plenty of gain combined with high definition. The original had lead and rhythm channels, as well as Resonance and Presence knobs. Though there is no Presence knob here, the Treble control offers plenty of high-end shaping capability. The original's Resonance knob has been labeled Bottom on the AmpliTube version, as it enhances the speaker bottom-end resonance. You can easily turn the Metal Lead V into a two-channel amp.

1. Set the gain at between 7 and 9 o'clock. This will provide a nice, crunchy rhythm that cleans up when you back off your guitar volume.
2. Now add a Diode Overdrive in the Stomp section. When engaged, it will quickly turn your rhythm "channel" into a lead one.

Metal Lead W

When it comes to metal players, it doesn't get much heavier or more aggressive than the late Dimebag Darrell. The Metal Lead W is based on a head that Randall Amplifiers built to satisfy the Pantera guitarist's thirst for in-your-face power. Its 300 watts of solid-state juice lent a potent punch to the riffs of many a purveyor of heavy music. A 3-band active EQ with midsweep channel and a 9-band graphic EQ allow extensive tone shaping. The original offered 16 digital effects, available here through similar effects in AmpliTube's Stomp and Rack sections.

The Metal Lead W offers fans of solid-state amps the immediacy of attack and the low-end tightness that transistors offer. Seven-string and drop-tuning guitarists especially seem to favor the controlled bottom of solid-state amplifiers.

As with any genre, metal has a subtle but wide variety of sounds associated with it. For the most part, though, it breaks down into classic and modern. The selective mid control, 9-band graphic EQ, and separate gain control on Metal Lead W allow you to experiment with these different sounds.

- A classic metal sound might have a bit less gain, a slight boost at 2 in the mids, and a flat-ish EQ, as pictured in Figure 5.8 (think: Black Sabbath).
- A modern metal tone usually has gobs of gain and a scooped EQ, sometimes called a *smiley face* (see Figure 5.9). The bass player would set his controls for more midrange to fit in the space carved out in the guitar sound (think: Napalm Death, Hatebreed).

Figure 5.8 Note the roll-off at 65 Hz on the graphic EQ to make space for the bass and bass drum.

Figure 5.9 A smiley-face EQ curve for modem metal.

British Amps

In the '60s, the British reconfigured American rock and roll, sending it back across the pond with some new twists. With American amps prohibitively expensive, the Brits developed their own amplifier industry, shaping the sound of future rock music.

British Blue Tube 30TB

For more than half a century, the Vox AC30 has been at the forefront of rock and pop music. It probably covers more styles than any other guitar amplifier. The Beatles and the British Invasion introduced its chiming tone to America, where Tom Petty and the

Heartbreakers and REM responded to its combination of grit and jangle. Back across the pond, Brian May of Queen was crafting the solo for "Bohemian Rhapsody" by driving the amp to the brink with a treble booster, and later the Edge added delay to forge U2's ringing anthems. Today, you're as likely to hear Hellacaster John Jorgenson using it for country as John Scofield employing it for jazz (see Figure 5.10).

Figure 5.10 The Vox AC30 combo.

The AC series actually began with the AC15, a 15-watt amp with a pair of EL84 tubes (or *valves*, as they were called in England) and a single 12″ speaker. The AC30 doubled the number of EL84s and eventually added a second speaker.

The AC15 was essentially the same circuit as the AC30, so you can approximate the earlier model by switching the AC30's 2×12 Open Vintage cabinet to a 1×12 Open Vintage model.

In AmpliTube this significantly darkens the sound. Diming out the Treble and Cut controls offers a lovely, warm British tone, but if it remains too dark, try switching to a 1×12 Open Modern cabinet for added brightness.

Soon after the AC30's initial release, a "Top Boost" circuit modification was offered, featuring an extra tube, as well as Treble and Bass controls. Its popularity led to its becoming a standard feature of the AC30. The combo with the two 12″ Celestions known as *Blue* or *Bulldogs* is especially prized and modeled here.

In an effort to compete with the ever larger and more boisterous crowds at the '60s rock concerts, Vox introduced the AC50, which provided the extra power inherent in a pair of EL34s run in a Class AB configuration. Switching the British Blue Tube 30TB's power section to 50W EL34 will provide you with the extra headroom and more open sound afforded by this high-powered model.

The tone layout on the British Blue Tube 30TB can be a little confusing. The control responsible for the upper "presence" is called the *Cut* control rather than a Boost. Unlike a Bright switch or Presence knob, it can really cut certain high-end frequencies. It is in juggling the Treble and Cut controls that the character of an AC30's tone is shaped.

Exercise 5

1. First, try setting the British Blue Tube 30TB With Bass at noon and the Treble and Cut controls full up (see Figure 5.11). This slicing sound might find its way into a mix, but more likely we need to trim the ear-splitting highs back a bit, especially if our guitar has single-coils.

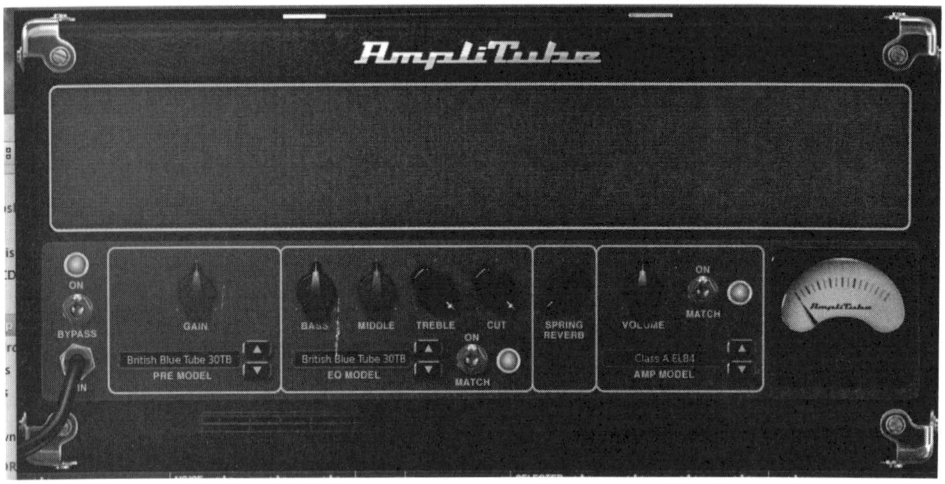

Figure 5.11 Full-on treble might be a bit too much.

2. Next, set the controls like those in Figure 5.12. We still have plenty of high end for cutting through the mix, but the top end is rounded off a bit.

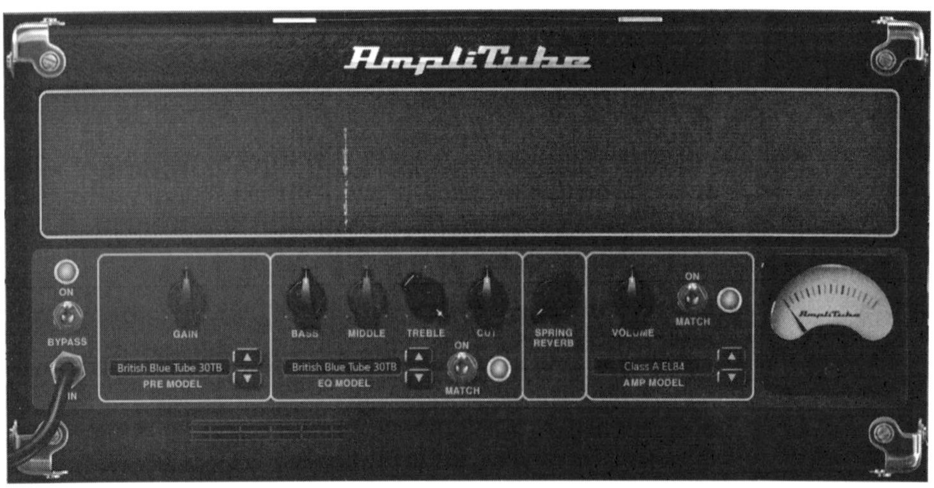

Figure 5.12 Rolling back the Cut control tames some of the high end.

3. Now roll the treble back to noon and turn the Cut knob fully clockwise (see Figure 5.13). This produces a more midrange-heavy tone, perfect for solos, especially when a Diode Overdrive is used to boost the gain.

Figure 5.13 This setting is excellent for stomp-driven solos.

British Copper 30TB

The British Copper 30TB harks back to the earliest days of the Vox AC30, when the Top Boost feature was an option rather than included. It also adds the famous Vox Vibrato/Tremolo circuit. (Here the two effects are labeled backward. Vib is actually the famous Vox bias modulation tremolo, whereas switching to Trem will yield a pitch warble like the old Magnatone amps—think: Lonnie Mack's "Memphis.") See Figure 5.14. The Tone control is authentically counterintuitive in that turning it up cuts the high end rather than adding it.

Figure 5.14 The Vibrato and Tremolo settings are labeled backward.

- The first AC30s used an EF86 preamp tube in the non–Top Boost channel. This tube provides a great distortion sound, evident here. Unfortunately, it was notoriously unstable, subject to failure from vibration, so Vox replaced it with a 12AX7.

Because vibration isn't an issue with a software version, you can enjoy the tone of the original AC30 without any reliability worries.

- The British Copper also defaults to the pre-Celestion speakers, used in the first 2×12" combos. Feel free to try the 2×12 Open Vintage model for the Blueback Celestions that were soon standard.
- The first 30-watt Vox amps came with a single 12" speaker, so if you want to retreat to the earliest Vox AC30 sound, you can switch off the Top Boost and swap out the 2×12 Gry British Vintage cabinet for the 1×12 Open Vintage one.

The British Blue Tube 30TB and British Copper 30TB, combined with all of the potential speaker and tube variations, and the Diode Overdrive will provide you with the full range of historic Vox AC sounds, from Beatle-esque chime to Brian May roar.

British OR

Unable to get manufacturers to sell him new amps for his fledgling London music store, Cliff Cooper was forced to design one himself. Fortunately, his electrical engineering background helped him make a good one. The amp's combination of blindingly bright orange-colored vynide covering and cartoon-like graphics instead of letter labeling for its unique controls gained his product instant notice. His Orange OR 120 head, modeled in AmpliTube, was soon appearing on stages with artists as disparate as Fleetwood Mac, T-Rex, Stevie Wonder, and James Brown.

Powered by a quartet of EL34s, the OR 120 can produce sounds that seem British or American depending on how its distinctive controls are set. The F.A.C. knob, represented by a high-pass filter icon, is a six-position selectable bass cut in the preamp. The first position to the left cuts the low frequencies a little, and each middle position cuts a little more, up to the last position clockwise, which cuts them considerably. Every position is musical, making this control perfect for placing your guitar in a mix. The bass clef labeled Hz adjusts the bass, while the treble clef indicated by kHz deals with the treble. The cryptically labeled HF.Drive is a kind of presence control. What the Sputnik icon above the Gain knob means is anybody's guess, but this actually sets the power-amp level.

- For a warm, American, Fender-style clean sound, roll the HF.Drive and Gain to about 9:30, set the F.A.C. control in the first position, and adjust the Bass and Treble to taste (see Figure 5.15).
- For more of a British, Vox-like tone, move the F.A.C. up to the third position, leave the Gain alone, and adjust the HF.Drive to sit at noon (see Figure 5.16).

The British Or model is ideal for tight, crunchy rhythms; classic rock leads; and clean tones unsullied by overuse. I find it to be one of the most dynamically responsive models in the whole AmpliTube catalog. It matches beautifully with its 4×12 British Or

Figure 5.15 These settings will produce an American clean tone.

Figure 5.16 Set here, the controls will yield a more British sound.

cabinet, but switching to other choices can help push its tone to one side of the ocean or the other.

British Lead S100

The loudness wars didn't start with CD mastering. In the '60s, when Pete Townshend, Jeff Beck, Jimi Hendrix, and Ritchie Blackmore wanted amps that "went to 11," they visited Jim Marshall, a former drummer who had begun building amps based on the Fender Bassman to sell in his music shop. His company's first effort, the legendary JTM 45, used a pair of KT66 tubes (the British equivalent of 6L6s). When Townshend found its 45 watts of power insufficient to make his guitar audible over the crashing of Keith Moon's drums, he encouraged Marshall to up the ante to 100 watts. Marshall added another pair of KT66s, and the historic 1959 100-watt Super Lead Plexi was born.

(The 1959 is a model number, not a date.) The British Lead S100 models the classic tone of this amp.

- Later versions of the Super Lead used EL34s, easily enough mimicked in AmpliTube by just exchanging the 100W EL34 for the 100W KT66 power stage.
- Want the original JTM 45? Swap out the 100W KT66 for the 50W 6L6 or 50W EL34 for earlier or later versions, respectively.

Eric Clapton's tone on the John Mayall & the Bluesbreakers record changed the sound of rock guitar forever and has become a touchstone for fans of the British blues sound. Clapton didn't want the typical JTM 45 head and bottom, preferring a combo that he could throw in the trunk of his car. Marshall built him a 2×12" version that became unofficially known as the Bluesbreakers model (see Figure 5.17).

Figure 5.17 The amp unofficially known as the Bluesbreakers model.

1. The first step toward achieving this tone is to substitute a 50W 6L6 power section for the British Lead's 100W KT66s.
2. Next, change the cabinet to a 2×12 Closed Vintage.
3. Finally, set the controls as pictured in Figure 5.18.

This will get you in the ballpark, but you'll have to wait for the effects chapter to flush out this classic tone's final secret.

Classic Marshall Tips Though associated with hard rock and metal, the early Marshalls were no gain monsters. The legendary rockers of the '60s and early '70s used them for clean tones and crunch rhythms, for the most part. At loud enough volumes, the power tubes would distort enough to provide a compressed sustain for solos, but most of the power and sustain of those classic rock leads came from sustain-laden Les Pauls combined with the powerful right-hand attack and left-hand vibrato of the era's legendary players. Solos might be enhanced through the use of "treble" boosters or fuzz tones, but by modern metal standards the early Marshall amps were relatively clean.

Figure 5.18 Adjust these settings to your instrument, pickups, and playing attack.

British Tube Lead 1

Introduced in 1981, the Marshall JCM800 series (named after James Charles Marshall's license plate number) instantly became the go-to amps for the decade's metal and hard-rock players. Though initially meant to be a revisiting of the classic Marshalls, the new series sported a brasher sound, channel switching, and master volumes, all of which appealed to a new generation of guitarists from Tom Morello to Kerry King. In America, some models were fitted with 6550 tubes—a sort of souped-up 6L6 —for an even bigger brasher tone. Though traditionalists balked, the JCM800 series became emblematic of the big-hair '80s.

- Feel free to goose the wattage up to 100W EL34's and experiment with 100 and 50W 6L6s to get a taste of the 6550 tone.

British Tube Lead 2

There arose among JCM800 owners in the '80s a trend toward hot-rodding their amps by increasing the gain and installing effects loops to handle rack-mounted modulation and ambient effects. Marshall took the hint, and the '90s saw the release of the JCM900 series. This version of the classic amp had a Gain knob that went to 20 in a nod to Spinal Tap's Nigel Tufnel's amps that went to 11. ("Well, it's one louder, innit?") AmpliTube's British Tube Lead 2's Gain knob stops at the normal 10, but its sound reflects the extra gain that made fans of Iron Maiden and ZZ Top's Billy Gibbons.

- The JCM900 series offered a wide range of heads and combo configurations that can be reproduced by adjusting the power section to various tube types and wattages.
- The purpose of an effects loop is to place certain effects, such as delay and chorus, after the preamp stage of an amplifier so that the modulation and ambient effects could be placed after the gain stage in the preamp. You can achieve a similar result by using AmpliTube's rack section for these effects.

Vintage Metal Lead

The demand for still more gain among metal players led Marshall to configure some of the preamp tubes in series so that they would overdrive each other. As happy as musicians might have been with the screaming sustain and massive volume of the resulting 100-watt 2203 models, some venue owners were less than thrilled with the earsplitting sound that resulted. This led to a need for an amp that could deliver the aggression and sustain of the 2203, but at lower volumes. Thus, the Marshall Master Volume series (MV) was born. Originally called the JCM 100, this head could simultaneously please both guitarist and club owner.

Of course, all of the AmpliTube models deliver their tone at any volume, but the Vintage Metal Lead emulates distinctively the edgy tone of this Marshall, as well as its capacity for gobs of gain and near infinite sustain—without the use of any pedals.

- If you want to keep the character of this amp while taming some of the shrillness, consider switching the cabinet to the 4×12 Closed 25 C cabinet.
- To emphasize the midrange frequencies, run the head through the 4×12 Closed 75 C.

Other

AmpliTube contains a number of amps that don't fall into the classic American or British categories but can be very useful for creating tones of your own. The majority of them are models created from scratch by IK Multimedia, offering some unique variations on the essential amp sounds.

Jazz Amp 120

Based on the Roland Jazz Chorus, this amp doesn't fall into the American or British category in part because it originated in Japan. It nevertheless has become its own kind of classic. Guitarists soon realized that the "Jazz" part was a suggestion rather than a limitation. While some jazz guitarists appreciated its ability to remain perfectly clean at higher volumes, the Jazz Chorus soon found favor with funk musicians who valued that same headroom. Pop guitarists fell for the huge sound of the built-in stereo chorus and vibrato. Though it was rare that anyone used its distortion feature, the amp responds well to overdrive and fuzz pedals. For a solid-state amplifier, it delivers a warm tone, which is one more factor that has kept it in production for 35 years.

- If you find the pure Jazz Amp 120 attack too hard-edged for you, try adding a little compression with one of AmpliTube's stomp compressors. (Using the rack compressor will not have the same effect, as it is after the amplifier in the chain.)
- Don't be afraid to experiment with other speaker configurations. Many will soften the solid-state bite of the amp, and some, like the 2×12 Open SL, will add a pleasant speaker distortion.

Custom Solid State Clean

IK Multimedia's Custom Solid State Clean model performs a similar function to the JC 120 in offering a clean tone with plenty of headroom. To these ears, the sound reflects the advances made in solid-state technology since the Roland amp was designed, offering a little more tube-like "give" in response to pick attack, making a compressor in front less necessary. Pushed hard, it can break up in a pleasing fashion that might create its own space in a mix when combined with an overdriven tube amp.

- You can soften the sound further or get closer to the bright, hard sound of the JC 120 depending on which cabinet you choose.

Music Man Amps After leaving the Fender Corporation, founder Leo Fender started a company called *Music Man* to continue manufacturing guitars and amplifiers. The Music Man amps were hybrids: The preamp section was solid state, and the power stage was tubes. Mark Knopfler, Eric Clapton, Albert Lee, and Johnny Winter, among others, used these amps.

As with all of the AmpliTube models, the Custom Solid State Clean allows you to swap out preamp and EQ models, so you might want to experiment with creating a classic Music Man amp (see Figure 5.19).

They came in a variety of wattage and speaker configurations; early models used 6CA7 power tubes, which were soon replaced by EL34s. Later models used 6L6s, so this gives you a wide range of power-amp choices—especially since some models delivered as much as 130 watts. As long as you stick to open-back cabinets for the combos, you're in the Music Man ballpark—though the company also offered closed-back cabinets for their heads.

Custom Solid State Lead

As with all great solid-state amps, the Custom Solid State Lead excels at delivering tight low end regardless of the speaker choice or configuration. This model offers a wide

Figure 5.19 By exchanging the Custom Solid State Clean's solid-state power amp for a tube one, you can create a model of a Music Man amp.

variety of gain and plenty of it. Its basic tone configuration of bass, treble, and presence keeps the essential character of the sound within tight parameters, but the plethora of potential cabinet combinations will provide plenty of coloration possibilities.

Custom Modern High Gain

Though displaying a fair amount of gain, this model is not really a modern metal monster—you will still likely need a pedal in front for metal solos. Its tone lies somewhere between classic British and modern American. Whatever you call it, it's a sound you may find yourself coming back to when you need an aggressive, but not too edgy tone.

Custom Solid State Fuzz

To quote the AmpliTube manual: "Based on AmpliTube 1's fuzz amp, this model is a very distinct and distorted sound, reminiscent of '60s fuzz sounds." IK is too modest: This model is the key to dozens of classic tones from golden age of Brit rock. I was going to wait until the effects section to cover this, but I got blindsided by this model, which reveals the secret of so many coveted rock sounds.

Remember when I mentioned in the British Lead S100 section that you would have to wait until the effects chapter to fully understand how to get the Bluesbreakers' tone? Well, this model gives it away. That legendary tone and many like it are impossible to nail down exactly without placing a pedal in front of the amplifier. No combination of vintage guitar and amplifier alone will accurately reproduce many of the solo sounds of the '60s.

Players such as Eric Clapton, Jeff Beck, and Jimmy Page would often place some kind of booster pedal between their guitar and the amp. Sometimes it was a fuzz, sometimes a "treble" booster—there were no overdrives like the Tube Screamer back then.

The names were often deceiving. A treble booster didn't just boost the treble, but it did accentuate it as the player moved up the neck. Fuzz pedals were sometimes "fuzzy" like the ones on "Satisfaction" by the Rolling Stones or "Over Under Sideways Down" by the Yardbirds, but often they were smoother, more overdrive-like, cleaning up nicely when the guitar volume was backed off.

The "fuzz" here doesn't sound like "Satisfaction" but more like the latter style. It is more of a booster than a fuzz, and, like a booster, it is quite dynamic, cleaning up when you lighten your guitar attack. It combines here with a Vox-type EQ stage and EL34 power section to conjure up the sound that influenced generations of guitarists, from Eric Johnson to Joe Bonamassa.

- Try the Bluesbreakers' settings from the British Lead S100 section but replace the S100 preamp with the Custom Solid State Fuzz pre model. To do this, you have to start with the Custom Solid State Fuzz and change the EQ to British Lead 1 and the 50W EL34s to 50W 6L6s. Remember to change the cabinet to the 2×12 Closed Vintage.
- When swapping other EQ sections while the Custom Solid State Fuzz is in the preamp section, you'll find that American EQs sound fizzy (bad), as opposed to fuzzy (good), with the exception of the American Tube Vintage and the Tube Vintage Combo. This makes sense, as the former (Fender Bassman) was the model for Marshalls, and the later (Supro) was used by Jimmy Page on early Led Zeppelin records.
- In fact, for early Zeppelin guitar tones, it is hard to beat the Custom Solid State Fuzz going into the Tube Vintage Combo EQ, the Single-Ended Class A powerstage, and either the 2×12 Open SL (Silvertone) or the 1×12 Open Vintage (Fender Deluxe) cabinet (see Figure 5.20).

Figure 5.20 Try this configuration for early Led Zeppelin sounds.

Summary Questions

You can find the answers to the Summary Questions in the Appendix at the end of this book.

1. What is the most important secret to getting great guitar tone?
2. What is the difference between American and British tone?
3. When does the power tube section of an amp have the greatest effect on the tone?

   A. At high volume?

   B. At low volume?
4. What is the tonal difference between Fender's Twin and a Dual Showman amps?
5. Which amp models emulate the sound on the first Led Zeppelin record?
6. Which model is best for the sound of bands such as Korn, Limp Bizkit, and Linkin Park?
7. Which set of tubes produces amore vintage sound in a Triple Rectifier model: 6L6s or EL34s?
8. Which amp's control plate features a Sputnik icon?
9. How do you create Eric Clapton's Bluesbreakers Marshall combo?
10. Which amp emulates Marshall's Master Volume series?
11. How do you construct a Music Man amp?

6 The Cabinet

In Chapter 5 you learned a little bit about how the cabinet and speakers can affect the sound of an amp. Let me emphasize that here—the cabinet and speaker combination through which an amp is sent can *enormously* affect the ultimate sound.

A few years ago, when I was the gear editor at the late *Guitar One* magazine, the Eminence speaker company sent us two 4×12″ cabinets, each with four different speakers in it, for a total of eight different Eminence speaker models. The company had rigged these cabinets with a floor switcher that looked like something from the telephone company in the '40s, with eight large, clunky mechanical switches—one for each speaker. The idea was to plug an amp head into the cabinets and turn on one speaker at a time, demonstrating the difference in sound—like the experience of going to the high-end audio store and trying out home-stereo speakers or studio monitors.

The difference from speaker to speaker was astounding. Though I've had speakers replaced before, the downtime between hearing the old and new speakers diminishes the effect; with the Eminence rig, it was instantaneous. Thanks to AmpliTube, you don't need that kind of massive setup to instantly hear how different an amp sounds when the speakers are changed.

The best way to learn about this is to experiment, experiment, experiment! But to point you in the right direction, in this chapter I'll touch on some of the qualities and uses of the 40 different guitar-amp cabinets offered in AmpliTube.

1×6 Small Combo

You might wonder why AmpliTube would want to include a cabinet based on an inexpensive solid-state Fender amp. We are talking about great tone, right?

I bring you back to context: In the right context, this kind of speaker may provide exactly the right tone. What is that context? Well, let's say you have a song that already has three or four guitar parts recorded: an acoustic strumming; a jangling arpeggio; and a full, round crunch rhythm part, doubled and panned left and right. You suddenly come up with a perfect syncopated, saturated rhythm that meshes perfectly with the crunch part and the jangly arpeggios. How do you fit it in the mix so that it carves out its own space without obscuring any of the parts that are already working—or the bass or the vocals?

This is where a 6″ speaker can come into play. This kind of speaker focuses on a narrow frequency range that leaves lots of room for other parts while allowing the new part to be distinctly heard. On its own it may not sound great, but in the mix it might prove to be just the thing.

1×10 Combo Modern

This cabinet is based on the 10″ Eminence Black Shadow speaker in the Mesa Boogie Subway Rocket combo. It provides a wider frequency response than the 6″ speaker but maintains some of the focus.

In a close-miking situation, there is little difference between miking a 1×10″ and a 4×10″ cabinet. When a 4×10″ is close-miked, the mics are often centered on the best sounding of the four speakers. It is only when more distant room mics come into play that the other speakers are heard. This is true for 1×12″ cabinets versus 2×12″ or 4×12″ as well.

4×10 Closed Modern

The '90s Marshall JCM410 cabinet on which this is based came with four 30-watt 10″ Celestion Heritage speakers. Lower-wattage Celestions are considered "warm" due to their tendency to mildly accentuate the lower midrange. They also soften the edge of higher midrange and eliminate piercing high end. This allows for crunchy rock tones and high gain solos that don't hurt the ears. They also cut off the bass frequencies well above the range of the typical electric bass, making for clarity in a band's live or recorded sound.

4×10 Open Vintage

This model is based on a 1967 Fender Super Reverb, which sported four Jensen speakers. As Jensens are not too different from Eminence speakers, this might be a good time to try an exercise.

Exercise 1

We're going to compare the sound of a 1×10″ cabinet to that of a 4×10″ cabinet first, with each exclusively close miked and then with some room mic added.

1. First choose your amplifier head and shut off any amp reverb. Make sure the room mic level in the Cabinet module is all the way off (counterclockwise).
2. Next pair the amp with the 1×10 Combo Modern.
3. Play and listen carefully.
4. Now click the right Cabinet Model arrow twice to replace the 1×10 with the 4×10 Open Vintage.

5. Listen again. You will not hear a radical difference in sound. Obviously the Jensen is voiced differently than the Eminence, but the overall sound does not reflect the presence of three other speakers.
6. Now set the room size for a Large Studio, dial the room level to 1 o'clock, and shut off Mic 2 so that the room sound is emphasized.
7. When you compare the speaker cabinets this time, you should hear the boxiness of the 1×10 accentuated, while the expansiveness of the 4×10 comes alive.

1×12 Combo

This cabinet is borrowed from Marshall's Valvestate 8010 combo. The Valvestate series were solid-state amps that employed one 12AX7 (also known as an ECC83) tube (or "valve" in Brit-speak) for their overdrive sound. This 1×12″ evidences an even frequency range and an open sound, less boxy than many small 1×12″ cabinets. Try Exercise 1 again.

1. Compare it to the 4×12″ Closed Vintage 1 to see how little difference there is when close miked.
2. Add some room mic and hear how the 4×12 fills the room.

1×12 MB II

Based on the Electro-Voice speaker-equipped cabinet of a Mesa Boogie Mark II, the 1×12 MBII offers ample evidence of how a speaker and a cabinet can alter the sound of an amp. For the purposes of this book, I'm comparing each speaker as it sounds being driven by AmpliTube's British Lead S100 amp model—essentially a JMT100. Through our first 1×12 Combo, much of the Marshall character is maintained—not surprising, as the 1×12 Combo is modeled after a speaker found in a Marshall amplifier. As soon as I switch to the 1×12 MB II, the sound takes on a distinct Boogie tone—not just any Boogie tone, but the one associated with early Santana, despite being driven by a classic Marshall head!

Celestion, Eminence, Jensen, Electro-Voice, and JBL These five speaker brands are found in the vast majority of classic and modern guitar amplifiers. Electro-Voice, JBL, Eminence, and Jensen are (or were) made in America, while Celestion is based in Great Britain.

Jensen Fender, Ampeg, Gibson, and other amplifier companies used Jensen speakers until Jensen stopped making them in the late 1960s. In the late 1990s, an Italian speaker manufacturer licensed the name and revived production. These speakers are lightweight and deliver a warm, even tone. They are not very efficient, but the speaker breakup that comes with driving them hard is part of their legend.

The first Fender Bassman and Bandmaster amplifiers were loaded with Jensen P15N Alnico speakers, while the Super Reverb was loaded with P10Rs. Alnico speakers are made with magnets containing aluminum, nickel, and cobalt. Cobalt is hard to find these days, contributing to the high cost of Alnico speakers.

Eminence Eminence began by making OEM speakers for other companies in Eminence, Kentucky, and still does. Later, the company began selling the speakers under the Eminence name as well. Today, the company makes a wide range of speakers, but early Eminence-based speakers produced a sound similar to Jensens.

Electro-Voice As amplifiers got bigger, the demand grew for speakers that could handle the greater wattage without breaking up. Players began replacing their Jensen and Eminence speakers with Electro-Voice (or EV) speakers. These speakers were able to deal with hundreds of watts of rock-and-roll power. When used with lower-wattage amps, they provide enhanced efficiency: An 18-watt amp run through an EV 12 will sound significantly louder than one run through a classic Jensen or a low-wattage Celestion.

The EV sound tends to focus in the midrange and low end, giving more bark than sparkle. And their heavy-duty magnets make them much heavier than Jensen, Eminence, or Celestion speakers—though this is obviously not an issue in AmpliTube.

JBL JBL (also marketed as Altec Lansing) speakers fulfilled a similar need to the EVs: increased power handling and efficiency if you were willing to deal with the added weight. Their frequency response sits higher than the Electro-Voice, making them prized by a certain generation of country guitar and pedal steel players for their ability to enhance "twang."

Celestion During the early '60s, British musicians discovered that the Celestion Blue speaker combined with an overloaded tube amplifier produced a glorious tone. The speaker's slower response smoothed distortion into creamy overtones, aiding sustain, while its rapid high-end roll-off reduced finger and pick noise. Matched with the Vox AC30 and Marshall guitar amplifiers, the Celestion Blue helped change the sound of rock and roll.

Lightweight and inefficient like the early Jensen and Eminence speakers, their slight midrange bump makes Celestions more suited to the Vox and Marshall grind, while the Jensen and Eminence seem better matched to the Fender sparkle. That said—and having put Celestions in both a Fender Twin and Fender Deluxe to great effect—as with all things related to guitar tone, it's whatever works for you.

1×12 MB III

Similar to the MB II, the MB III's tone is less specifically Boogie/Santana. A little less midrange focused, it still retains that Mesa Boogie 1×12″ sound, highly suited to singing solos.

1×12 Open Modern

Based on a VHT Pitbull 1×12″ combo, this cabinet comes with one of the modern Eminence speakers and offers a more aggressive upper midrange than the EV or Celestion cabinets. It is well suited for focused hard-rock tones.

1×12 Open Vintage

This is based on a cabinet shape that is my personal favorite: the 1965 Fender Deluxe. Technically, this is the Fender Deluxe Reverb, which is less cube-like than many 1×12″ cabinets and combos. There is something about this shape, wider than it is high, that seems to add an open quality to the sound regardless of the speaker involved. The modeled version in AmpliTube is loaded with a model of the original Jensen that Fender employed.

To demonstrate that, er, size matters, this might be a good place to check out the Size knob in the Cabinet model section (see Figure 6.1).

Figure 6.1 The Size knob lets you experiment with changing the cabinet size for different tones.

Exercise 2

1. Choose the 1×12 Open Modern cabinet that comes just before this one.
2. Alternate playing through the 1×12 Open Modern and the 1×12 Open Vintage.
3. Turn the Size knob on the 1×12 Open Modern to about 2:30.
4. Now alternate playing through the 1×12 Open Modern and the 1×12 Open Vintage again. Hear how the sound of the 1×12 Open Modern now bears more resemblance to the 1×12 Open Vintage than it did before?

This will allow you to get some of the Deluxe cabinet openness out of the different speakers in each of the 1×12 cabinets in AmpliTube.

2×12 Closed Vintage

This model of the Marshall 1922 2×12″ cabinet with Celestion Greenbacks is perfect for turning the British Lead S100 into a classic Bluesbreakers combo. It can also help demonstrate how not all amp configurations are compatible with all speaker setups.

Exercise 3

1. Call up an American Vintage T amplifier and listen to how it sounds with its matching cabinet: the 2×12 Open T J120.
2. Switch the cabinet to the 2×12 Closed Vintage.
3. Note how there seems to be a phasing issue that creates a subtle but unpleasant honk in the midrange.

Of course, this cabinet will match up well with a variety of other amps, and there is always the chance that you happen to love the sound with the Vintage T. If so, go for it.

2×12 Gry British Vintage

This it the cabinet from the Coppertop Vox and models the Goodmans Audiom 60 speakers that Vox used before switching to Celestions. Just as the 1×12 MB II will nudge any amplifier toward the Boogie/Santana sound, this cabinet will edge any amp with which it is paired toward the classic early-'60s Vox tonality.

2×12 Open Vintage

Within a short time, Vox decided to upgrade the speakers in its new AC30 to Celestions. The ones that the company used were the famous "Blues" modeled in AmpliTube. These Alnico speakers are prized among guitarists for their dampened attack, warm lows, mellow upper-mids, and chiming top-end.

- Find the default 2×12 Open T J120 speakers matched with the American Tube Clean 1 too piercing? This cabinet is just the thing to tame those knife-like highs.
- The 2×12 Open Vintage also works for domesticating some Marshall amps, while retaining a distinctly British character.

2×12 Open TJ120

Remember those countrified JBLs from the sidebar earlier in this chapter? The D120 model comes paired with the American Tube Clean 1's Twin emulation. Although it may seem as if I have been anxious to replace these speakers with other, mellower ones, they definitely have their place. There is a reason why they were so popular with Fender owners for so long.

- For jazz, their high-end definition ensures that dark archtops never become muddy.
- A certain era's ice-pick country tone is unobtainable with any other speakers.

- Pop purveyors will love how they help retain sparkle and clarity when running guitars through a multitude of effects.
- Finally, funk fanatics will like being able to chunk clean rhythms that cut through horns, keyboards, and background vocalists.

2×12 JP Jazz

This cabinet is modeled on the Roland Jazz Chorus' two 12″ speakers. Though you probably won't want to use this cabinet with any kind of distorted head, you might be surprised to find that it adds sparkle to many tube heads, thinning out their clean tone in musical ways that just might fit your mix.

2×12 Open Vintage

This 2×12″ Sears Silvertone model originally sported a pair of Jensen speakers. The cheap housing definitely has an effect on the tone. You will find that it lends an interesting lo-fi character to any amplifier you match with it.

4×12 British Or

This eye-catching cabinet comes with four Celestion Vintage 30s.

Here is an exercise to demonstrate how varied the same speakers can sound in different cabinets. The shape, the density of the wood, and the design can all conspire to make the same amp run through the same four speakers produce a different tone.

Exercise 4

1. Call up a British Lead S100 and match it with a 4×12 British Or cabinet.
2. Record a few bars of chords onto the Four Track Audio Player/Recorder's Track 1.
3. Click on Track 2 and add an instance of AmpliTube with the same head. Make sure that the reverb is turned off on both track heads and that the settings are the same.
4. Run the Track 2 head through a 4×12 Metal T 1. This cabinet contains the same speakers as the 4×12 British Or.
5. Now mute Track 1 and record the same chords onto Track 2.
6. By muting each track, you can listen to the sound of the same part through the same head and speakers, but employing a different cabinet.
7. The difference is subtler than using a different type or number of speakers, but it should be distinctly audible.

4×12 Closed 25 C

Four Celestion 25-watt Greenbacks in a Marshall cabinet re-create another classic sound of the '60s. These low-wattage speakers break up beautifully when pushed. Their frequency response provides a warm midrange without a trace of edge or "honk." This cabinet is ideal if you want to play loud, distorted guitar, but you don't want to make anyone's ears bleed. It is a sound that glories in being given plenty of space in a mix—think AC/DC or Free. If you want something that will cut through a crowd, try the...

4×12 Closed 75 C

The Celestion G12T-75 was developed for the higher gain amps of post-'60s rock. It exhibits a more focused and aggressive midrange, while retaining the vintage Celestion warmth. It is easy to compare to the Greenback. Just toggle back and forth with the cabinet arrows. Still not cutting enough? Check out AmpliTube's next cabinet...

4×12 Closed J120

We discussed the properties of JBL speakers back in the 2×12 section and the sidebar. If you're going to play scooped midrange metal, there is no point in emphasizing the mids with a Celestion-equipped cabinet; the JBL120F speakers offer a more even frequency response, equally usable for jangly and crispy clean tones and the aforementioned metal. These heavy-duty and heavyweight speakers are no longer made, but through the magic of AmpliTube are easily available for recorded or live use—and you don't have to carry the massive load that four of them in one cabinet represents.

4×12 Closed Modern 1

This is a replica of the Marshall 1982A cabinet that originally sported Celestion G-12T-75-watt speakers. This cabinet works well for recording a modern metal sound.

4×12 Closed Modern 2

This cabinet is modeled on the Mesa Boogie 4×12 Rectifier using Celestion Vintage 30 speakers. It sounds brighter and cleaner than the Orange cabinet with the same speakers.

4×12 Closed Vintage 1

This cabinet is modeled after a design that Jim Marshall came up with solely because he thought it looked good. He didn't like the way two straight-front 4×12s looked when stacked, so he slanted the one to go on top. It was only later that he realized that the slanted top helped projection in big halls. In closer quarters, the slant will angle two of the speakers up at the player's ear, allowing lower volume while still allowing the player to hear the guitar over the bass and drums.

Of course, neither the design nor the throw direction of the speakers means much when recording. Though there is no doubt a subtle sonic difference results from the shape of a slanted or angled cabinet compared to a straight one, the more obvious tonal difference here, as compared to the Closed Vintage 2 modeled on a straight Marshall cabinet, is due to the speakers. Vintage 1 sports '70s Celestion Greenbacks, notable for their heavier magnets and stiffer cones. This cabinet emphasizes a higher range of mids than the older Greenback-equipped Vintage 2.

4×12 Closed Vintage 2

This model of a Greenback equipped "straight" Marshall 4×12 rocks a mid-'60s sound, with lower, softer mids. Its frequency range complements the Vintage 1 nicely, and using them together will create a huge sound.

2×15 Closed B J130 and D J130

Here is a stellar example of how a different cabinet can affect the sound of the same speakers. Both of these 2×15 cabs sport JBL D130F speakers, but you will find that the B J130, representing a '60s Fender Bassman bottom, and the D J130, from an early Fender Dual Showman, sound significantly different. The Bassman bottom is brighter and tighter, while the Showman bottom has a definite high-end roll-off combined with a little woof in the low end.

Both cabs are cool for bottom-string twang, baritone guitar work, and Texas blues rhythm played on heavy strings. There is a reason why Stevie Ray Vaughan often used 15″ speakers.

Rotary

Due to the radical effect that rotary speakers have on the sound of the guitar, I have placed the discussion of their unique tone in Chapter 8, "Effects."

AmpliTube 2 Cabinets

- 4×12 Metal F1 and 2: Fender MH 4×12 head equipped with Celestion G-12K-100 speakers
- 4×12 Metal T 1, 2, and 3: Mesa 4×12 rectifier cab with Celestion Vintage 30 speakers
- 4×12 Metal V 1, 2, and 3: Peavey 5150 4×12 cab with custom Peavey speakers
- 4×12 Modern M 1, 2, and 3: Marshall JCM800 4×12 cab with Celestion G-12T-75 speakers
- 4×12 Vintage M1 and 2: Marshall vintage 4×12 slanted cab equipped with Greenback Celestions

These cabinets represent models that were held over from AmpliTube 2 into AmpliTube 3, so that those who upgrade to 3 could recall their presets from AmpliTube 2. AmpliTube 2 did not have the sophisticated miking available in 3, so each number refers to a different miking technique (different mics or miking a different speaker within the same cabinet) used on the same cabinet. These cabinets are not just a convenience for those who upgrade, though; they offer a wealth of tone-shaping possibilities as an adjunct to all the other speakers. Moving the AmpliTube 3 mics around just adds to the fun.

Summary Questions

You can find the answers to the Summary Questions in the Appendix at the end of this book.

1. What is one reason why you would want to use a 6″ speaker cabinet?
2. True or false: There is very little difference between the sound of the same speaker close miked in a 1×12″ cabinet and a 4×12″ cabinet.
3. What sort of sound would be best produced by a 25-watt Celestion-equipped cabinet?
4. What sort of sound would be best produced by a 120-watt JBL-equipped cabinet?
5. True or false: Any speaker cabinet can sound good with any amp configuration.
6. True or false: With the same speakers loaded, a miked straight 4×12″ Marshall cabinet will sound very similar to a slanted version of the same cabinet.
7. What are two styles of playing that go well with a 2×15″ cabinet?
8. What type of music favors the Celestion G-12-75 speakers?

7 Miking

The guitar sounds that you love most are often ones you have heard on record—that classic album you play until the grooves wear out or the MP3 you put on repeat to hear over and over again as you learn the licks.

That sound is made up of more than the guitar, amp, cabinet, speaker, and effects employed by your idol; it includes the type of microphone(s) used to record that guitar and amp and those effects. Moreover, it consists of the exact position the aforementioned mic(s) occupied in the room relative to the speakers. It also may incorporate the sound of the room itself, whether a small, dead, carpeted space or a large wooden and glass room.

The best way to learn how to mike up your amp is by trial and error. As I have said elsewhere, with AmpliTube you have the advantage of not having to run back and forth from the control room to the amp room every time you want to move or change a mic. But there is still a lot of experimenting to do, and here's why:

- Small changes in the position can result in large changes in the sound.
- Different mics will reveal different frequency ranges.
- When you start combining mics and adding room mics, it gets even more complicated.
- Using multiple cabinets and/or amps exponentially raises the complexity ante.

It may seem overwhelming at first, but getting a handle on how to mike your amplifiers is the key to turning good recorded sound into great recorded sound.

Some basic principles to keep in mind include the following:

- Moving the mic closer to the amp provides more definition, increased highs and lows, and less room sound.
- Pulling the mic back will cause the sound to become less detailed, more "midrange-y," and more blended with the ambience.

Figure 7.1 A 4×12 cabinet miked in the studio.

- Pointing the mic at the center of the cone will yield more active highs and better transient detail but fewer lows.
- Moving the mic toward the outer rim of the speaker increases the low frequencies, resulting in a warmer, softer, or more powerful sound.

Important: The "right" microphone is the one that produces the sound you seek. Learning how each of these mics affects the sound coming out of the amplifier is key to producing a killer track.

One Microphone

Using just one microphone to mike a cabinet is the simplest way to do it. Still, the variations in tone can be considerable.

Exercise 1

The choice of guitar, amp, effects, and routing is only part of the equation. The microphone determines a significant portion of the tone. To get an idea of the proportion that each factor plays, try this:

1. Grab a guitar and choose an AmpliTube rig. Use a clean head, a 1×12 bottom, and Module/Rig Selector 1 to keep things simple.
2. Now go to the cabinet module (Cab A). If there are two mics, mute Mic 2 and make sure Mic 1 is a Dynamic 57.
3. Turn the Room Level all the way off and shut off any reverb on the amp.
4. Start with the mic dead center of the circular shadow in the middle of the cabinet (see Figure 7.2). Play and listen.

Figure 7.2 The microphone should start here.

5. Now grab the mic with the cursor and slide it to the left, all the way to the outer circle shadow, moving the cursor up a little to make sure that you don't draw the mic away from the amp (see Figure 7.3). Listen again.

Figure 7.3 The microphone should end up here.

You should be able to hear easily the difference in tone between the two positions. Now try starting again at the center and moving the mic a quarter of an inch at a time toward the outside rim of the speaker. You may not hear the difference at first, but keep doing it and listening carefully, and you'll eventually be able to discern slight changes in the quality of the sound as you move the mic away from the center.

By grabbing the mic and moving your cursor up and down, you can move the mic closer to and farther from the speaker, respectively. The effect of this will vary depending on the microphone type. With the Dynamic 57, you will hear the proximity effect that is a feature of the Shure SM57 on which it is based. A *proximity effect* means that as the mic gets closer to the source, in addition to the volume getting louder, the low end increases. Some other mics don't react like that; they maintain the same tone regardless of distance.

This brings us to the next experiment of substituting different mics in the same position. Let's go through the models that AmpliTube offers.

A Short Microphone Primer A microphone uses a mechanical transducer to convert sound into an electrical signal. The transducer element of a microphone is called its *element* or *capsule*. Types of microphones are often identified by their transducer—that is, condenser, dynamic, and so on.

The Dynamic Microphone This type of mic tends to be tough and inexpensive. It can also withstand considerable volume before being damaged, making it ideal for recording loud rock guitars. Dynamic mics tend to emphasize midrange, which is often desirable when miking guitars. A dynamic mic operates essentially the reverse of how a speaker works. A small movable coil is positioned in the magnetic field of a permanent magnet and attached to the diaphragm. Sound waves enter through the windscreen and move the diaphragm. The vibrating diaphragm moves the coil in a magnetic field, creating a varying current in the coil through a process called *electromagnetic induction*.

The Condenser Microphone Also sometimes called a *capacitor microphone* or an *electrostatic microphone*, the diaphragm here acts as one plate of a capacitor, with the sound vibrations creating changes in the distance between the two plates. As only a very small mass must be moved by the sound wave, condenser microphones are extremely sensitive and can produce a high-quality audio signal. In the real world, they require a power source, provided either as phantom power or from a small battery. Condenser mics capture a wider range of frequencies than most dynamic mics, making them ideal for accurately representing certain types of chime and twang. They also combine well with dynamic mics to fill in the frequencies that dynamic microphones lack.

The Ribbon Microphone Ribbon microphones employ a thin metal ribbon suspended in a magnetic field. The ribbon is wired to the microphone's output, and its vibration within the magnetic field generates the electrical signal. Because the ribbon mic is open to sound both front and back, it detects sound in a bidirectional (also called *figure-eight*) pattern. It offers high side rejection, which can be an advantage in some miking situations. (This is more useful in a real studio than a virtual one.) Steve Albini used Coles 4038s to capture Kurt Cobain's electric guitars on Nirvana's *In Utero,* but older ribbon microphones are relatively fragile, thus unsuited for high-volume guitar recording. Current materials make modern ribbon mics more robust and improve the effective dynamic range at low frequencies. This combination has made them all the rage with 21st-century guitarists. These microphones can now be safely used to capture more of the actual sound of the guitar amplifier than dynamic mics, without the fizz that can result when you're using condensers too close to the speaker-cone center.

Dynamic 57

This is based on the Shure SM57 (see Figure 7.4), a small-diaphragm dynamic microphone that has long been the musician's workhorse; it is affordable and hard to break, and it sounds great. Some all-time classic guitar tones have been achieved with nothing more than a 57 properly placed in front of a great-sounding guitar, through a great-sounding amp, played by a great player.

Figure 7.4 The classic Shure SM57.

The model here mimics the original's proximity effect, so if you move it farther away from the speaker, the low end will diminish. This is not necessarily a bad thing. If your mix sounds muddy, you can reduce the low end of an amp miked with a Dynamic 57 by backing the mic away from the speaker, reducing bottom without changing the essential quality of the sound.

Dynamic 20

This model of the Electro-Voice RE20 re-creates a more even response across a wider range of frequencies than the 57. It is especially appealing in the low end, making the Dynamic 20 also a great selection for electric bass.

Exercise 2

Here is a good way to get a quick picture of how different two mics can sound:

1. Set up a rig the same way as you did in Exercise 1—in other words, a 1×12 cab with one Dynamic 57 model, reverb off, and so on.
2. Position the mic in the center of the speaker and play.
3. Now click on the right arrow next to MIC Model (see Figure 7.5). This will call up the Dynamic 20. Play again. You will hear how the sound gets mellower and less aggressive than it does with the Dynamic 57.
4. Click the left arrow and go back to the 57. Go back and forth playing through each one to get a good sense of the difference in sound.

Figure 7.5 Click back and forth on these arrows to compare the two mics.

Vintage Dynamic 20

This is modeled on the AKG D 20, a vintage dynamic large-diaphragm microphone often used to mike bass amps and kick drums during the '60s. Warm and present, it has a pleasant midrange presence and a smooth high-frequency roll-off. It can be a little dull by itself, but it works great in conjunction with a 57 to add focused low end.

Dynamic 421

The Sennheiser MD 421's (see Figure 7.6) boosted hi-mid frequencies make it ideal for capturing aggressive, driven tones. That, combined with an ability to handle high sound-pressure levels, makes it the second most popular dynamic mic for guitar amps. In the real world, the fact that the Shure 57 is a third of the price helps keep it on top.

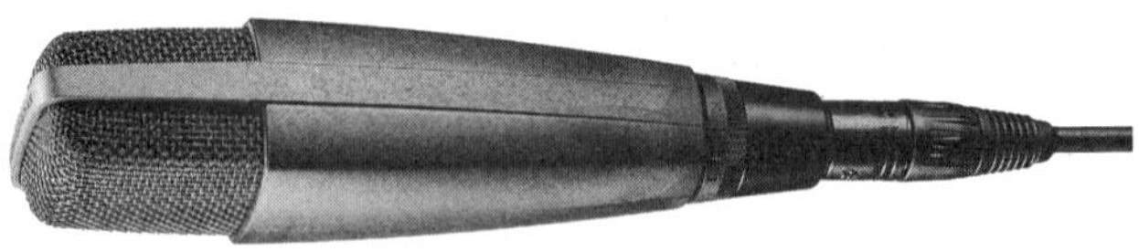

Figure 7.6 Sennheiser MD 421.

Dynamic 441

Boosted mids are not always desirable in guitar tone. The flatter frequency response Sennheiser 441 (see Figure 7.7), on which this is based, enhances sparkly, clean tones.

Figure 7.7 Sennheiser 441.

Dynamic 609

The Sennheiser e 609's (see Figure 7.8) construction allows it to be hung over the front of a cabinet, making it easy to mike up amplifiers. It can be very midrange-y, but you may want this, or you can temper it by combining it with a warmer condenser microphone.

Figure 7.8 Sennheiser e 609.

Condenser 12

This mic model is based on a legendary tube recording microphone: the AKG C 12 VR condenser. It offers extremely detailed high frequencies, so be careful, as it can sometimes emphasize a speaker's fizziness when facing the cone's center.

Exercise 3

Here is a way to A/B two microphones if, unlike the 57 and the Dynamic 20, they are not sequential in the MIC Model list:

1. Set up a rig like the one in Exercise 1.
2. Make Mic 1 the first mic you want to try and Mic 2 the one to which you wish to compare it. Place the mics side by side in front of the speaker cone (see Figure 7.9).

Figure 7.9 Place Mic 1 and Mic 2 like so.

3. Now, with both mics engaged, you can slide the MIC Blend slider (see Figure 7.10) all the way from exclusively Mic 1 to exclusively Mic 2 , allowing you to quickly A/B the two microphones.

Figure 7.10 Use the Mic Blend slider to compare the two microphones.

Condenser 67

This mic model is based on the tube-powered Neumann Model U 67 (see Figure 7.11). Shortly after its release in 1960, this mic became ubiquitous in recording studios, thanks to its solid build and wide variety of applications from guitars to strings to vocals. Consider using it in conjunction with a 57; blending in a little will round off the bottom and soften the highs.

Figure 7.11 Neumann Model U 67.

Condenser 84

The Neumann Model KM 84, a great small-diaphragm condenser mic, modeled here in the Condenser 84, is the standard for small-diaphragm mics. Though it has a wider frequency response than the 57, it has a little more mid focus than the Condenser 67.

Condenser 87

This is a model of the Neumann Model U 87 (see Figure 7.12), another extremely popular large-diaphragm mic. It sounds similar to the 67 with perhaps a bit more top and a tighter low end. Keep in mind that the differences between certain mics will be more pronounced with some rigs than with others.

Condenser 170

This is a model of the Neumann TLM 170 R/ST (see Figure 7.13), a high-quality large-diaphragm condenser microphone used in studios worldwide. It was one of the first microphones with transformerless technology. Neumann claims its sound is more transparent than the character-filled 67 and 87. The model here seems to focus on lower mids in contrast to the wider frequencies of the 67 and 87 or the higher mids of the 57.

Condenser 414

The Condenser 414 is modeled after the large-diaphragm condenser AKG C 414 XLS (see Figure 7.14). It is a flat and clean/open microphone that is often used for acoustic

Figure 7.12 Neumann Model U 87.

Figure 7.13 Neumann TLM 170 R/ST.

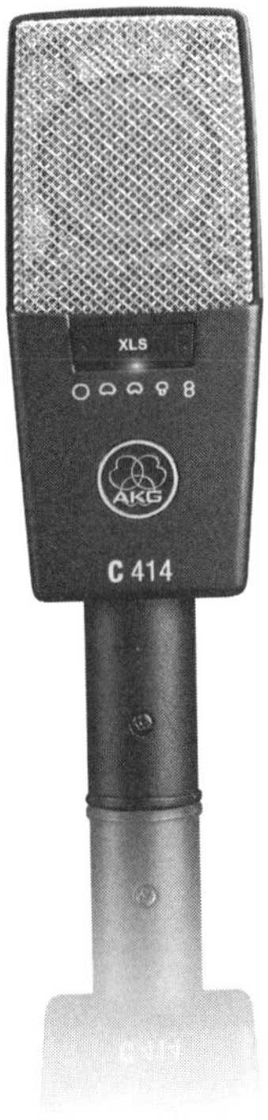

Figure 7.14 AKG C 414 XLS.

instruments. It works great when combined with another mic, backing off the 414 to capture the air and using a 57 or a 121 for the close work.

Ribbon 121

Released in 1998, the Royer R-121 ribbon microphone was the first ribbon mic able to handle close-miking duties on electric guitar. The Royer R-121 ribbon microphone has been creeping up on the Shure SM57 recently as the go-to electric guitar mic. While the

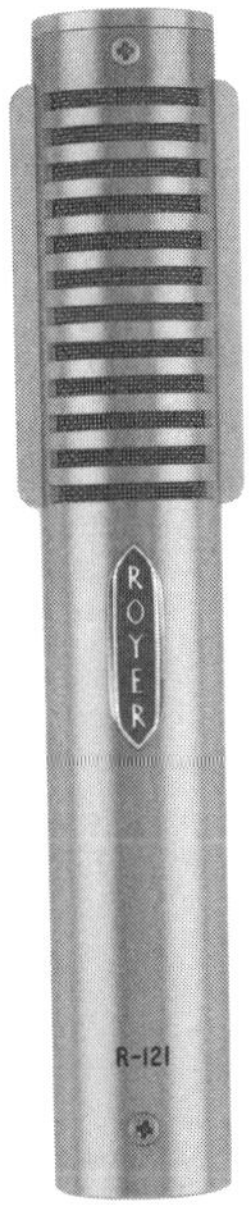

Figure 7.15 Royer R-121 ribbon mic.

57 still offers classic rock-and-roll aggression, the 121 tends to capture the full sound of the particular amp being used. The Ribbon 121 reproduces the original's present midrange and smooth roll-off on the high frequencies, making it ideal when you've got too much "fizz" on high gain and metal tones. Using this mic will reduce the unwanted high frequencies without reducing presence.

Ribbon 160

The Beyerdynamic M 160 is a double-ribbon hypercardioid microphone, popular for acoustic applications. It has a focused midrange and is great alone or for warming up a 57.

Velo-8

In the 1930s, ribbon mics were called *velocity microphones*. The Velo-8 models the Groove Tubes VELO-8 ribbon microphone. This is a double-ribbon figure-eight microphone with a unique neodymium element. This model offers tons of low-end girth, mixing very well with other microphones to create full-bodied recordings. Combine this with the Tube Vintage combo for great classic blues tone.

Two or More Microphones

As you can see, this miking stuff can get complicated. AmpliTube offers you 15 different choices, and each one will sound different depending on dozens of minor increments of movement left and right or forward and back. This is all before adding another microphone and/or another cabinet with its one or two mics or before turning up the room mics and deciding on the size of the room and the width of the room mic stereo spacing. Whew!

Let's just take it a step at a time. After you have spent some time (and by some, I mean a lot) trying the different mics, A/B-ing them, and moving them around, it's time to add a second microphone to the equation. Remember to keep working with the same amp sound, the same guitar, and no reverb so that your experiments will reflect only the changes in mics and mic arrangement.

One thing we begin to deal with when using two mics instead of one is *phasing*. Sound is produced in waves that resemble Figure 7.16. When two mics are placed in different locations, the waves of certain frequencies can reach the mics at different points along their peaks and valleys. When these waves combine on the recording or coming out of the monitor speakers, the valleys of one cancel all or part of the peaks of the other, affecting the sound.

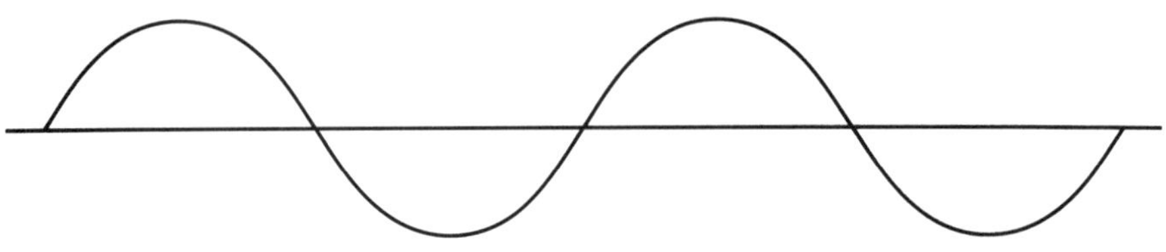

Figure 7.16 Sound waves.

If you look at the right side of the Cabinet/Microphone module of AmpliTube, you will see three buttons. One (M) is the Mute button for the selected mic. Below that is the (S) Solo button for that mic, and the circle with a line through it below the Solo button is the Phase switch (see Figure 7.17).

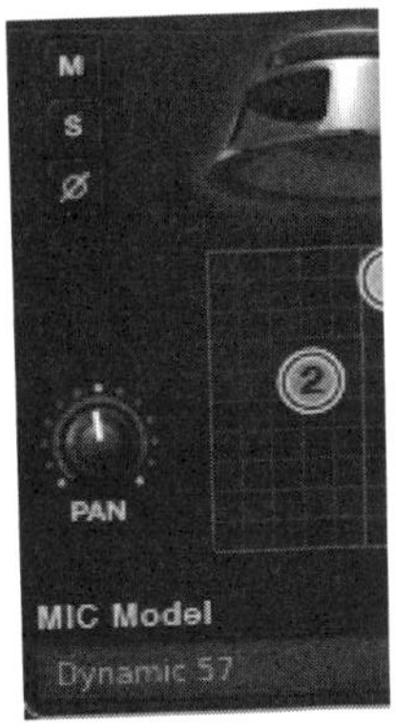

Figure 7.17 The Phase button is the one that looks like a circle with a line through it.

When you have chosen the two mics that you want to use and positioned them where you think you might want them, you might hear the sound thin out rather than get fuller. If this is not the effect you are seeking, try clicking on the Phase button for one of the mics. It doesn't matter which one; just don't click both, as that will be the same place you started. If the sound is better, leave the phase switched; if it is worse, don't. If neither position suits you, you will have to keep moving the mics around until their relative positions achieve the sound you are looking for, which may or may not involve reversing the phase of one of the mics. You can check the pair for phase cancellation by panning them to the same spot and listening in mono. If they are out of phase, the volume and tone quality will decrease significantly.

In general, dynamic mics work better for close miking, while condensers are better for adding air. Ribbon mics may work for either. As with most things involving recorded sound, these are starting places, but try anything you like—if it sounds good to you, it's fine.

Room Miking

The wealth of options for the two cabinet mics offered in AmpliTube will give you a sound that is as full or as thin as you like. You can pull them back to add a little air or glue them right up against the cabinet to put the sound directly in your face. But AmpliTube doesn't stop there. Sometimes you might want an ambience more like a live performance in a club or a hall, where the sound bounces off the walls. Sure, you could add digital reverb, but that never sounds quite the same as a real room.

AmpliTube's Room feature (see Figure 7.18) is based on impulse recording, which means that when you increase the level of the room ambience, you are hearing actual recordings of a real room, miked at varying distances from a sound source, added to the sound of your model. This type of processing requires a lot of CPU power, so it's not available in economy mode.

Figure 7.18 Here is where you set the size, width, panning, and amount of room ambience.

There are two basic ways to approach the room sound:

- One method is to hold off adding this ambience until the mix, so as to add just the right amount to make the part sit nicely among the other instruments and vocals.
- The other method is to choose an ambience for the guitar first and then build the other sounds around it.

If you think you want a very "roomy" sound, I advise going with the second method. A distant guitar can be very atmospheric and evocative, but if the other parts of the mix—drums, bass, vocals—aren't treated in a sympathetic manner, it can either get lost or stand out, as if on a different recording. Also remember to check the phasing with the closer mics by clicking on the slashed-circle symbol in the Room module.

Mic Panning

While the room mics can help you place the guitar in the foreground or background of a mix, panning allows you to set the sound wherever you like along the left-to-right stereo spectrum, even when using a single amp and cabinet routing. It can also help the guitar sound bigger or smaller in the mix.

How Does Panning Work? A panning knob essentially sets the volume of sound coming out of each speaker. When you turn it to the left, it increases the volume in the left speaker and decreases the volume in the right one. Turning it to the right does the opposite. Keeping the panning knob in the center makes the volume of the guitar sound in both speakers equal, which our ears hear as the guitar coming down the center.

In the simplest case—one amp, one cabinet, one mic, no room, no rack effects—you can select the panning location with the Pan control in the Mic section or in the Selected Module section of AmpliTube's GUI.

Master Pan One thing to keep in mind is that the Selected Module Pan control can override any mic panning that you might do. For example, you might pan one or more cabinet mics hard right. If the Selected Module Pan control is in the center or to the right, you will hear the sound coming out of the right speaker, but if you pan the Selected Module left, the sound will come out of the left side regardless of how any mics, including room mics, are panned.

When you start adding mics, it gets more complex, but the sonic and spatial possibilities are exciting. We have discussed how multiple mics—cabinet and room—can affect the sound; now it's time to explore how you can use miking to locate the sound in space.

Tighten Up

If you want to keep the sound pinpointed in the spectrum, make sure that both the cabinet mics and the room mics, if used at all, are panned to the same degree. You should write down the pan value showing in the Value box next to the Selected Module section and apply the same value to all microphones. If you're using multiple amps and/or cabinets to shape the sound, make sure they, too, are panned to the same location. If it is a mix with many instrument and vocal parts, it's often a good idea to keep each guitar track tightly located. Even in a sparse mix, it's sometimes cool to have guitar parts panned hard left and right with no crossing over, leaving a huge space for vocals, bass, and drums.

Simple Spread

One way to make your guitar sound a little bigger is to use two mics and pan them slightly apart—you'll be amazed by the sonic difference. To hear this clearly, once again, make sure the room mics are off and the amp is devoid of reverb.

You can spread them on either side of the center, or you can have one mic hard right or left and the other a little less so. Either way, your guitar will take up more sonic space, so be sure there is room for it in the mix.

Room to Grow

Another way to increase your instrument's sonic footprint is to pan the close mics to one side and the room mics to the other side. This is a subtler effect than panning a rack delay to the other side.

Exercise 4

Here is an exercise to illustrate this effect:

1. Set up a single rig and pan both cabinet speakers hard right. Keep the room mics and amp reverb off.

2. Now add some room mics panned hard left. Lower the width of the room mics to keep the sound focused on the left (see Figure 7.19).

Figure 7.19 Pan the room opposite to the cabinet speakers. Note that the room mics are close together.

Note how this maintains a sense of the guitar coming out of the right speaker but adds to its presence in the stereo spectrum as well.

You can use this trick with two rigs as well, to place a little of each side's character on the opposite side.

Exercise 5

When your have separate rigs panned hard left and right, this is an exercise that should help you hear the difference between having some room panned opposite or not.

1. Set up two rigs in the Module/Rig Preset 2 configuration. Use whichever clean rigs you like for each.
2. Pan both of Cabinet A's mics to one side and both of Cabinet B's mics to the other. Turn off any room mics and amp reverb.
3. Set up Rig A's Stomp section like you see in Figure 7.20. The delay will help separate the spread, and the filter will give that side its own character.
4. Add an Octa-V to Rig B. You should now hear the rigs panned hard left and right, with lots of space in the middle.
5. Now add a small room miking setup to each side, panned hard in the opposite direction. That is, pan the right-side rig's room to the left and the left side's to the right.

Figure 7.20 Add a Step Filter and a Delay set thusly. The delay time is 30ms.

6. You will still get a sense of each distinct rig panned left and right, but they will take up a little more room in the mix.

Remember, too, that panning can affect phase relationships, so be sure to check that panned guitars are in phase.

This is just the beginning in terms of panning your guitar. When we get to Chapter 9, "Routing," the panning possibilities will increase yet again. Just remember, sometimes a single pinpointed guitar sound can be the most powerful.

Summary Questions

You can find the answers to the Summary Questions in the Appendix at the end of this book.

1. To significantly change the sound of a recorded guitar, you must move the mic a minimum of:

   A. Two feet

   B. One foot

   C. One inch

2. What is the result of moving a mic closer to the cabinet?

3. What is the result of moving a mic closer to the cone of the cabinet?
4. What are the three types of microphones commonly used to mike guitar cabinets?
5. What is proximity effect?
6. Which dynamic mic does *not* boost the midrange frequencies?
7. Which ribbon mic has become very popular for close-miking electric guitar?
8. What do you need to be careful about when using two or more mics?
9. Which quality modes allow room miking in AmpliTube?
10. How does panning work?

8 Effects

There are many electric guitarists who are quite happy plugging their instrument directly into an amp and leaving it at that—you will rarely see a traditional blues guitarist with a pedalboard. Every guitarist should strive for the best possible basic tone (before adding any effects) and learn all of the possible sounds that can be wrested from solely the instrument and the amp. There is a wealth of sonic possibilities to be mined from using a multi-pickup guitar's various positions, picking closer or farther from the bridge and even behind the nut, manipulating tone controls and whammy bars, and so on. Required viewing for anyone attempting to play the electric guitar is *Jeff Beck: Performing This Week...Live at Ronnie Scott's*, an exhibition of the tonal variety contained in one guitar, one amp, and the fingers of a genius.

That said—effects are fun! One of the advantages of electric guitar over other instruments is how well it takes to everything from subtle coloration to massive mangling by modifiers housed in stompboxes, racks, and computers.

Ready for another rule? Here goes: When you are using effects, most of the time, less is more—except when more is more. If that sounds contradictory, it isn't quite. Most of the time a small amount of distortion, reverb, delay, chorus, compression, and so on will do the trick. I will explain the reason for this as we discuss individual effects, but it is always a good place to start. There are those times, however, when what you seek is to make the guitar unrecognizable as such; in those instances, by all means lay it on thick.

Most effects fall into categories, so let's go through them and get an idea of what each one does and how it can help you record great guitar tones. When it comes to effects—as in many other areas of life—categories can be flexible. I will roughly follow the ones chosen by AmpliTube to make them easier to find in the software, merely changing the order, as often it is important to understand one effect before you can understand another.

Dynamics

In music and recording, the word *dynamics* refers to control over the volume of the signal. Dynamic range is the distance between the softest sounds produced and the loudest. Control of the dynamic range is important in all music but is especially crucial with recorded music for a number of reasons:

- Sudden loud peaks can distort mics, inputs, tape, or DAWs in an unwanted way.
- If the volume of the audio on a single track is uneven, the soft parts will get lost in the mix, and the loud parts will stick out.
- When the dynamic range is reduced, soft parts become audible, loud parts are controlled, and often the whole mix can be made louder. The downside of this is that certain kinds of musical expression are lost.

Compression One of the most effective ways that musicians control dynamics is with *compression*. A compressor is basically an automatic volume control. It rides herd on the level of the guitar signal put through it. It can make loud signals (caused by hard pick attack) softer and soft signals (caused by tentative picking) louder. This could be done by continuously adjusting a volume pedal or a fader on a mixing board, but a compressor is more efficient.

Whether stomp or rack version, a compressor is basically a variable gain amplifier that acts as an automatic volume pedal. Whether you strike the strings firmly or softly, the compressor keeps the volume at a preset level. When the string vibrations slow and the volume drops, the compressor's amplifier keeps raising the volume in an attempt to maintain the same level as the initial attack.

Even though the two stomp compressors in AmpliTube sport only two controls, it is beneficial to know some of the possible parameters of compression to better understand the differences between the two models here and to get a better grasp of the controls on the rack version.

- Threshold sets the input volume level at which compression kicks in. Any signal lower than the threshold is passed through without being compressed.
- Signals above the threshold are made quieter by the amount specified by the Ratio parameter. A compression ratio of 4:1 means that if a signal becomes louder than the threshold by 4 dB (decibels), the compressor will allow an output of only 1 dB.
- Attack and Release control how quickly the compressor reacts to input-level changes. Attack determines how quickly the compressor clamps down after a signal exceeds the threshold. Release determines how long the compressor takes to return to normal operation when the signal falls below the threshold. Longer attack times let pick attack come through, while shorter times create a squashing, almost reverse effect.

Stompbox compressors can be used for various purposes.

- They can increase "sustain." I put quotes around "sustain" because compression increases only apparent sustain; that is, the actual note doesn't last longer—it just appears to do so because the amplifier in the compressor keeps raising the volume as it decays. If you were to time the points where the sound disappears completely, with and without compression they would be identical. The difference is a compressed note stays loud longer, seeming to sustain more.

- Compression can even out arpeggios. Whether the notes are flatpicked or fingerpicked, it is the rare guitarist who can make every note exactly even in volume—a compressor can help.
- Some country players like the way a compressor with a fast attack will make the note sound as if it is being swelled in with a volume pedal, a common pedal steel effect.
- By placing a compressor in front of a clean amp, you can use it to boost solos without adding distortion.
- Put in front of an overdrive, a distortion pedal, or a distorted amp, it can increase sustain while allowing you to use less gain on the pedal or amp. This makes for more articulate solos.

Compressor

This AmpliTube stomp compressor is modeled after one of the high-end boutique versions, such as the Keeley or Empress pedals. It is quiet and offers a subtle compression with a slow attack, which is great for adding the aforementioned boost or evening out pick attack. The Comp level controls the amount of compression, which in this case means that it is simultaneously adjusting Attack, Release, and Threshold amounts. Level controls the output, allowing you to keep the level identical when on or off or to add as much as 15 dB boost.

Dcomp

Here AmpliTube has modeled the classic MXR Dyna Comp (see Figure 8.1). Cheap and noisy, it nevertheless became a fixture on country picker pedalboards through the '70s and '80s. They enjoyed the fast attack that took the edge off Tele through Twin solos and offered up pedal steel–style swells on chords. Andy Summers of the Police loved the way it brought out the chime in his chorused arpeggios, and Adrian Belew employed it both for tight, clean funk comping and to keep his animal noise effects sustaining. The Sensitivity knob performs the same function as Compressor's Comp knob, while its Output controls the level.

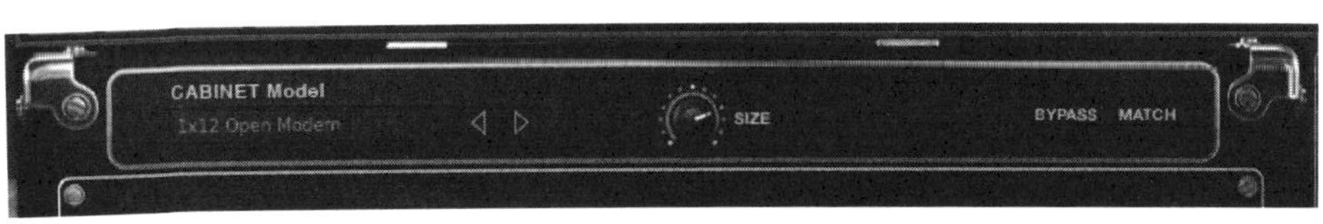

Figure 8.1 The MXR Dyna Comp was the first compressor available in stompbox form.

Tip: Do the following exercises using an amp set for a clean sound without any breakup.

Exercise 1

1. Place the AmpliTube Compressor in stomp slot 1 and turn it off.
2. Play eighth notes on the open low E string, playing softer, then harder, then softer again. Note how the volume gets louder and then softer.
3. Turn the Comp knob on the Compressor all the way up and the Level to 9 o'clock and then repeat the picking experiment. Note how the volume doesn't get louder or softer.

Exercise 2

1. Place the Compressor in stomp slot 1 and turn it off.
2. Add Dcomp to stomp slot 2 and turn it off.
3. Set the Compressor Level at 9 o'clock and the Comp all the way up.
4. Set the Dcomp output at noon and the Sensitivity all the way up.
5. Now go back and forth, turning one on and the other off, to compare the sounds and effects.

Exercise 3

1. Leave the compressors in place but off and place an Overscream in stomp slot 3.
2. Set the Overscream for Tone at noon or below, Drive at 10 o'clock, and Level at 2 o'clock.
3. Play through the Overscream alone first and then add the compressors one at a time.
4. Note how each, in its own way, adds sustain without adding gain, allowing fast single-note runs to remain distinct and articulate.

Other

If it seems strange to be getting into the "other" section so soon, it is because the effects contained there also deal with dynamics. So before we leave that topic, here are some "other" ways that AmpliTube lets you deal with level.

Volume

AmpliTube provides a virtual volume pedal that you can control with a MIDI controller. A volume pedal lets you be constantly in command of the level that you are sending to the amp—without taking your hands off the guitar. Check out the knob marked Min in the pedal's upper-left corner. This lets you set a minimum volume with your heel all the way back—say, for rhythm. This way, you can't accidentally shut your volume all the way off.

Some volume pedal tips:

- The volume pedal can be placed before or after overdrives and distortions. Before means that as you back it off, you will lower the signal to the overdrive, often cleaning up the sound. Placing it after allows you to have the same amount of overdrive or distortion at any volume.
- Usually the volume pedal is placed before ambient effects, such as delay and reverb, so that when you turn the volume down, the delay and reverb tails continue.
- Swelling the volume pedal gradually from off to louder can create beautiful chordal pads, especially when combined with delay and sometimes chorus.

Step Slicer

As extreme an effect as this might seem, it is actually just another form of volume control. (Technically, it could fall under modulation, but to make it easier to find in AmpliTube, we'll cover it here.) It lets you make the volume louder and softer in rhythmic increments that can be synced to a clock. The shifts in volume can be radical, creating synth-like effects, or mild for more tremolo-type sounds. You can tap the tempo in if you are playing it live with a drummer, or you can sync it to the clock of a DAW.

Some Slicer tips:

- Distortion effects work better before and modulation effects work better after the Slicer.
- Adding a compressor, either alone or with distortion, will allow notes and chords to ring all the way through long Slicer patterns at slower tempos.

Swell

Perhaps you prefer to walk around the stage or studio while doing volume swells, or you don't have a MIDI controller for AmpliTube. This is where Swell comes in. It will automatically swell your volume without your having to manipulate a pedal.

Some Swell tips:

- To make Swell work properly, be sure that each note you play is finished ringing before you play the next one, or Swell will not trigger properly. You can ensure this by damping the note with your picking hand or by lifting the finger playing it while keeping it on the string before you play the next note.

- Don't try to play rapid runs while using Swell—you can't damp that fast, and that is not what it's for.
- The same rules apply as with a volume pedal when it comes to location in the chain.

EQ

With all the tone controls on your guitar, the pedals, and the amp, you might wonder what you need EQ for. That's a good question. EQ is one of those things that is best applied sparingly. If you find that to get the sound you want, you need radical EQ from a pedal or rack unit, you might want to consider changing guitars, pickups, amps, cabinets, mics, or mic position first.

That said, EQ pedals are a way of quickly changing from one tone to another without having to do any of the above. Graphic EQs are called such because when you adjust the individual frequency bands, you get a graphic representation of your tone—you can actually see the peaks and valleys of the adjustments. Each slider represents a frequency range—for example, the 100 Hz slider may affect up to 150 Hz, while the 200 Hz might go down as low as 151 Hz and up to 250 Hz. More bands can mean a narrower range affected by each and/or a higher and lower frequency range overall.

7-Band Graphic

AmpliTube's 7-Band Graphic EQ covers the guitar-friendly territory between 100 Hz and 6.4 kHz. The Level slider allows you to make sure that boosting or cutting frequencies doesn't necessarily mean boosting or cutting volume.

- When you are entering the world of recording, it's a good idea to learn what the different frequency ranges sound like. Spend some time boosting and cutting each of the seven bands and listen to the effect.
- A stompbox graphic EQ can help turn a single-coil pickup into a humbucker sound for one or two tunes. Try setting the sliders the same as in Figure 8.2 and play a single-coil guitar through the Overscream. Then turn on the EQ and check out how it sounds more like a humbucker-equipped one—and the Overdrive responds in kind.

10-Band Graphic

"Well, it has three more." (Imagine a Nigel Tufnel voice.) The 10-Band Graphic deals with a significantly larger frequency range, extending from 31 Hz to 16 kHz. "Dimebag" Darrell used one like it to get his early sound with Pantera.

If an acoustic guitar isn't handy, try putting a "smiley face" curve into the EQ and strum away on the neck pickup (see Figure 8.3).

Figure 8.2 Note the curve on the EQ: perfect for turning a single-coil sound into a humbucker one.

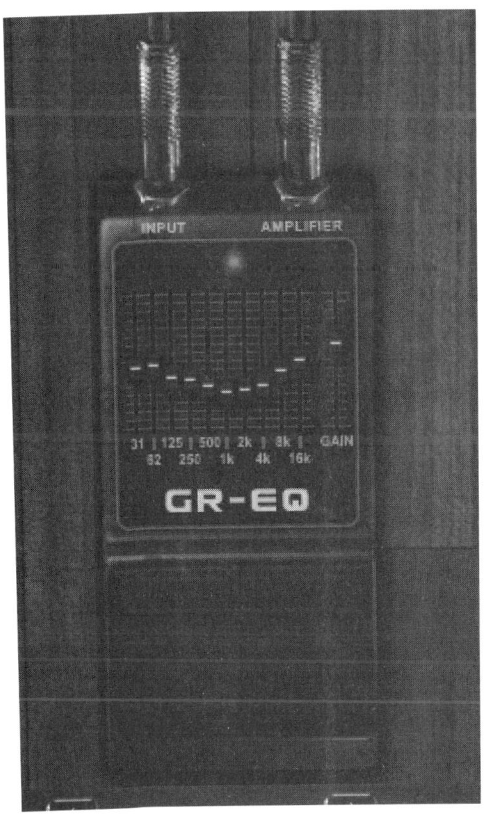

Figure 8.3 This EQ curve is sometimes called a "smiley face."

Filter

Filters and EQs are similar: They both deal with the frequencies that comprise the tone of your sound. The tone controls on your guitar and amp are a kind of filter. The difference is that when we refer to filter effects, we are often talking about movement, where the frequency being filtered is swept up and down in some fashion. EQs, on the other hand, are static—set to one place and left there. A quick lesson in some filter terminology is in order here.

- A high-pass filter lets only the sound above a selected frequency through—essentially allowing just the treble to pass.
- A low-pass filter lets only the bass through—that is, the frequencies below the chosen point.
- A band-pass filter lets you choose the frequency to be let through, silencing almost all the frequencies on either side of it.

Envelope Filter

An envelope filter is like a wah pedal in that it sweeps through frequencies. But, rather than responding to a moving foot, it does so automatically in response to the amount of input that it sees. Hit your guitar string softly, and the sweep is less than if you hit it firmly. Softer hits will not produce as much treble as harder hits. Envelope filters can be a very expressive effect on solos if you vary your pick attack. This effect also works well for funky rhythms. A little distortion placed after the effect helps bring out the frequency sweep.

LFO Filter

AmpliTube's LFO Filter pedal is similar to an Auto Wah—in both, the frequency is continually swept up and down by an LFO, or *low-frequency oscillator*. The LFO Filter can offer a welcome change from tremolo or chorus, to spice up recordings and give them a more modern sound.

Exercise 4

You can create a cool, pulsating synth bass track by combining the LFO Filter with some other stomps and AmpliTube's Automation feature (see Figure 8.4).

1. From left to right, load a Compressor, Octav, LFO Filter, and Distortion into the stomp module.
2. Set the Compression Comp all the way up to keep the notes going and to send an even signal to the octave pedal for better tracking.
3. Set the Octav as shown, with some original signal as part of the sound to bring out the upper frequencies.

4. Set the LFO Filter as shown and automate the Freq parameter in your DAW to move slightly up and down over time but *not* with the beat. This gives some motion to the part to keep it interesting.
5. Add some distortion, and you have a synth bass part with no keyboard skills necessary.

Figure 8.4 Create a pulsating synth bass with this setup.

LFOs Low-frequency oscillators are used in tremolo, chorus, flanging, phasing, and other effects, so you should understand their basic concept.

An oscillator produces a signal in waves, with peaks and valleys. How closely these peaks and valleys are scrunched together is called the *frequency*. A low-frequency oscillator usually puts out a, well, low frequency that is not sent to an audio output but is used only to modify another signal. Here it modifies the bass-to-treble response of the LFO Filter. By adjusting the distance between the LFO's peaks and valleys, we change the speed of the effect; by changing the level of the LFO, we change the depth.

Rezo

This stomp adds chiming overtones to the notes you play. All the controls are interactive, so the best way to get to know what it can do is to work with one fader at a time. To make this easy, record an arpeggiated chord into AmpliTube's 4-track audio player/recorder or your DAW. This will leave your hands free to play with the different parameters. The Rezo can supply everything from bell to sitar-like sounds, but it can easily get out of hand, so watch the output level. Also, you might want to place a rack compressor after it, as well as a stomp compressor before it.

Step Filter

It is easiest to think of this device as similar to the Step Slicer, except that instead of drawing in volume patterns, you draw in your own tonal shift patterns. Another way to view it is as an LFO Filter that lets you determine exactly when the frequency shifts up and down. The Step Filter is another effect that can offer you many of the textures enjoyed by keyboard and guitar synthesizer users without your having to learn a new instrument or install a synth pickup in your guitar.

- There is no way to blend a dry signal in with the effect, so this is a good place for parallel routing, which we will cover in the next chapter.
- Be sure to play with the destination control (Dest), which determines how much of the step programming controls the frequency cutoff and how much modifies the resonance. Some radical resonance mods can create rhythmic sounds that don't overpower the mix.

Wah-Wah

The wah-wah pedal is essentially a foot-operated, active filter. A battery or an adapter powers a small amp circuit that keeps your volume up as you sweep from bass (heel) to treble (toe) frequencies. The differences between wah-wah pedals come down to construction and voicing, with some offering control over the latter.

AmpliTube offers four flavors of wah-wah pedal, three of which offer an Auto setting, which actually turns the pedal into an envelope filter rather than an auto wah. This is handy for trying out the different tonalities if you don't have a MIDI controller handy. All four can be automated to any wah rhythm you choose in your DAW.

Wah pedals can be used rhythmically, controlled by foot, envelope, or automation, for expressive vocal-type tonal shifts during solos. Or, they can be used like the filter sweep in a synthesizer to slowly shift the frequency of arpeggios or sustained chords.

Here is a quick look at some of the characteristics of each wah-wah pedal:

- **Wah.** The IK Multimedia wah-wah has the classic wah sound. With the highest output of the four pedals, it will distort all but the cleanest amps.
- **Wah 10.** This is the Ibanez wah favored by John Frusciante of the Red Hot Chili Peppers. It has a depth control to let you modify the amount of the wah effect and the most even response of the four. It is the only one without an auto function. This wah also offers a setting for bass players.

- **Wah 46.** Based on the Vox V846 pedal made in Italy by Thomas Organ, this model was favored by Jimi Hendrix.
- **Wah 47.** Purportedly an exact reissue of the 846, the Vox 847, modeled here, sounds different—otherwise, why include it? This model has the most vocal sound of the four.

Caution: You may come across a lot of heated debate on Internet forums about which is the best-sounding wah-wah. The best-sounding wah—and for that matter, the best-sounding anything—is the one that sounds best to you.

Distortion

Distortion is one of the most confusing terms in music. When you are talking about recording guitars, bass, drums, vocals, horns, or strings, it might be a bad thing. Too, you normally don't want to push solid-state mixing boards of computer DAWs into distortion. On the other hand, should you be recording through a vintage tube mixing board and pushing the limits of 2" tape, a little distortion is what makes this old-school setup sound so good.

When it comes to distortion pedals, it gets much more confusing. To be precise, overdrive pedals and fuzz tones are also distortion pedals, as they both distort the guitar signal. Without getting technical, the difference between overdrive, distortion, and fuzz is, for the most part, a matter of degree and tone quality.

- An overdrive pedal tends to deliver a smooth sound and a minimal amount of distortion, and it sounds most like a real amp overdriving, cleaning up nicely when the guitar volume is backed off.
- A distortion pedal normally offers more gain and an edgier sound, and it may not clean up when you lower your guitar output.
- A fuzz pedal usually produces a fizzier, less focused distortion, and it may or may not add an upper-octave harmonic or clean up when the guitar volume is lowered.

To add to the confusion, there is a lot of overlap, with overdrives offering distortion and sometimes fuzz, and fuzzes that sound as smooth as overdrives. The important thing is not what category the pedal belongs in, but how it sounds and whether that sound works for your music.

BigPig

A perfect example of the aforementioned confusion is IK placing their model of what many would consider one of the most famous fuzz pedals of all time, the Electro-Harmonix Big Muff (see Figure 8.5), in the distortion rather than the fuzz section.

No matter, the model here delivers all the enormous harmonic richness and near-infinite sustain of the original. If you are looking to create Frippertronics soundscapes or White Stripes grunge, seek no further.

Figure 8.5 The E-H Big Muff offers near-endless sustain.

Crusher

Until recently, the sound of digital distortion was considered a complete no-no. The influence of electronic music on guitarists has changed that. Nine Inch Nails fans will love the way the Decim (for decimate) control on this pedal lowers the sample rate, creating a sound once considered grating and now considered cool—when you think about it, much like the story of classic distortion in the '50s.

- Modulating the Res control (actually the cutoff frequency of the low-pass filter) with a MIDI expression pedal or automation can create many interesting effects.
- If you find the sound more grating than cool, try putting it in front of a more typical distortion source, such as another pedal or an overdriven amp, to add a bit of a modern edge.

Diode Overdrive

Another recent trend in distortion is the rediscovery of boost pedals. As I have said, the secret to many classic guitar tones is a vintage amp kicked into high gear by a treble booster. The modern equivalent is the raft of session guitarists nudging their amps with boosters from companies such as Keeley and Xotic. Too, if you are going to use an amp, whether vintage or boutique, for the great tone that it produces when distorted, why place an overdrive or distortion pedal in front of it that will color that sound?

The Diode Overdrive handles this kind of boost brilliantly. Use it to push a Vox AC30 model for classic British Invasion tone or to kick a Super Reverb model into blues badness. The trick is to keep the Dist knob way down low and just drive the front end of the amp with the necessary amount of Level. The Diode Overdrive offers relatively transparent boost that maintains the character of whatever amp you use, whether it is modeling a small Silvertone or a metal stack.

Distortion

Like the Boss DS-1 that it was modeled after, AmpliTube's Distortion stomp evidences a dynamic response to pick attack: Its raspy distortion softens with a softer stroke and gets edgier when you hit it harder. For the most obvious dynamics, keep the Dist control at noon or below.

Feedback

This stomp simulates the Boss Super Feedbacker & Distortion, which uses an oscillator, to re-create whatever note you are playing when you hit the FBK (Feedback) button. This simulates the effect of feedback, though the note won't change in response to finger vibrato or whammy-bar action. An Overtone knob controls the octave of the "feedback" note.

The infinite sustain of this note offers all sorts of possibilities in terms of adding effects, such as the Step Slicer or Filter, and changing the overtone through MIDI control or automation.

Metal Distortion

The Boss Metal Zone pedal is its biggest seller of all time. This is not because so many famous metal players use it—for the most part, they don't (though some non-metal artists have found creative uses for it). Modern metal pros more often get their tone out of many of the amps modeled here, with maybe the aid of a booster drive—and so can you.

The Metal Zone's enormous popularity is more likely because thousands of young would-be metal players can't afford the Rectifier stack used by their hero. For them, this pedal beautifully emulates their idol's screaming tone through whatever amp they or their parents can afford—at a volume said parents can approve of. Its comprehensive EQ control, modeled here, lets them and you dial in a massive metal sound, shaped to taste.

Because you can get this tone with the actual amps here, the most likely uses of this stomp would be to quickly go from a clean Fender tone to a mighty metal rasp or some new type of tone that has nothing to do with typical metal. Avoid adding the pedal to already high gain amps, or your tone will lose all definition while adding much unwanted fizz.

Metal Distortion 2

The pedal that preceded the Metal Zone at Boss was the HM-2 Heavy Metal pedal. Though it delivers a dark, classic metal sound, reminiscent of Sabbath, Judas Priest, et al., it, too, was employed less by actual metal players than by alternative and noise bands, such as Swervedriver, My Bloody Valentine, Curve, and the Jesus and Mary Chain.

This is a great pedal for the sounds of angst. Just be aware that while the L knob in the Color section increases the lows, the H knob increases the mids, not highs. For a brighter sound, you'll want to back off the H level.

Overdrive

With all the major manufacturer and boutique overdrives that have come since, it is easy to forget how good the old SD-1 Boss Super Overdrive sounds. The AmpliTube model captures its ability to deliver singing solo sustain without tons of gain. It sounds great through anything—from a clean Jazz Chorus to a rampaging Rectifier.

The SD-1 adds a bit of its own color and a fair amount of compression, but it remains a solid workhorse three decades after its introduction.

OverScream

Modeled on the Ibanez Tube Screamer TS9 (see Figure 8.6), this is the go-to overdrive against which a host of boutique pedals are judged. Stevie Ray Vaughan put it on the map, and modern blues players swear by it, but it is adaptable to any kind of music. Its smooth drive will give you plenty of slightly compressed sustain if you crank it, but to get the most out of the pedal, you should keep the drive at 10 o'clock and dig in, dragging the sustain out of your fingers and guitar—like Stevie Ray.

Figure 8.6 Stevie Ray Vaughan made the Tube Screamer the blues-approved pedal.

Pro Drive

The Pro Drive is modeled on the Pro Co Rat. The Rat's decades of appeal lie in its ability to go from subtle, smooth overdrive favored by jazz artists such as Bill Frisell,

John Scofield, and Kurt Rosenwinkel, to nasty, edgy distortion in the hands of a Jeff Beck or Thurston Moore (see Figure 8.7). Turn down the distortion, and it delivers dynamic drive that cleans up nicely when you back off the guitar volume. For jazzier tones, you'll want to turn down the treble, which on the Rat means counterintuitively turning the tone clockwise.

Figure 8.7 Jeff Beck pulls a plethora of sounds out of the Pro Co Rat.

The Ambass'dor

Though it is nowhere near as well known, you might think of the Marshall Guv'nor pedal upon which the Ambass'dor is based as the British version of the Pro Co Rat. Just as the Rat provides the full range of classic American overdrive and distortion, the Guv'nor was good for every level of vintage Marshall dirt from the subtle crunch of the JTM45 to the more aggressive grind of the 800 series. Both pedals are extremely musical and work best in front of a clean-ish amp.

Fuzz

The popularity of fuzz skipped a couple of decades. It was common in the '60s, featured on hits such as the Rolling Stones' "Satisfaction" and the Yardbirds' "Heart Full of Soul," as well as playing a large part in the tonal excursions of one James Marshall Hendrix. But soon guitar amplifiers began to sound better and better when breaking up, master volumes allowed guitarists to get distortion at any volume, and pedals started to sound more like a real amp breaking up. The combination of those things and the general tendency toward faddishness of popular music sounds conspired to put fuzz tones on the back burner—until the '90s, when alternative bands such as Sonic Youth, My Bloody Valentine, and more recently Radiohead on the one hand, and tone fanatics such as Eric Johnson and Michael Landau on the other, rediscovered the joys of fuzz.

Another reason fuzz may have fallen out of favor is that it is not easy to get a good sound out of many types of fuzz. One reason for this is that the original pedals were finicky, with the same transistors sounding different from pedal to pedal. Of course, these differences are part of what makes fuzz fun: Each pedal and type of pedal has its own character. Another issue is that there are certain tricks to getting great tone out of a fuzz pedal. Some are pedal-specific; I will deal with these as we go from fuzz to fuzz in AmpliTube. Others, like the following, are more general:

- The sound of a plucked note dies out rather quickly on many fuzz pedals if your attack is soft or tentative. In general, these effects respond best to a hard, aggressive pick attack.
- Some fuzzes sound best through a clean amp. These pedals tend to be compressed and have a lot of sustain of their own; any additional distortion will further compress and muddy the sound.
- Some fuzz pedals are fizzy and thin-sounding and have very little sustain. Why would you want such a sound? This type of fuzz is great for adding aggressive cut when combined with the smoother distortion of some amps and pedals.
- Many vintage fuzz pedals and their modern clones require that you place them first in the effects chain. (See the sidebar at the end of this chapter.)

It is great for the world of sonic variety that fuzz is back. Fat fuzz can fill out a mix, while thin fizzy fuzz can cut through one. AmpliTube 3 offers eight different flavors, with one to fit almost any musical recipe (see Figure 8.8).

Figure 8.8 The flavors of fuzz.

Fuzz Age and Fuzz Age 2

This effect is modeled after the Arbiter Fuzz Face, a classic germanium-transistor fuzz stomp effect. The first Fuzz Face arrived in 1966. The original electronics were contained in a circular-shaped metal housing, modeled after a microphone stand's cast-iron base. The arrangement of controls on the original suggested a face, with controls as eyes, the switch as the nose, and the logo as a mouth. The original Fuzz Face was discontinued in 1974 or 1975. A short-lived reissue was made by Crest Audio in 1976, who revived it again from 1986 until 1990. In 1993, Dunlop Manufacturing, Inc. took over production (see Figure 8.9). In the late 1990s, Arbiter reissued the pedal as well.

Figure 8.9 The Dunlop Fuzz Face is an accurate version of the original, without many of its attendant problems.

The pedal contains only 11 electronic components. In the '60s, component tolerances were not as tightly controlled as they are today; two circuits utilizing the same components might sound very different. Germanium transistors also tended to be sensitive to temperature, with the sound changing as the equipment heated up. It was common to go through up to 20 Fuzz Face units to find a really good one, as no two were the same. The hardware reissues have essentially solved these problems, and, being software, these issues do not apply in AmpliTube.

AmpliTube 3 offers two stomps modeling the Arbiter Electronics Ltd. Fuzz Face, both labeled Fuzz Age: a red one and a silver one (Fuzz Age 2). The red version came originally with AmpliTube Jimi Hendrix; the silver one was added in AmpliTube 3. As I mentioned, the Fuzz Face pedals could sound radically different from unit to unit. This was not necessarily a bad thing; though some pedals sounded bad, others just sounded different. The silver version of the Fuzz Age emphasizes the low end, while cutting off most of the fizzy highs, whereas the red version is a more full-frequency model, without the midrange emphasis of the Class Fuzz. Both of these demonstrate a looser, fuzzier tone than the Class Fuzz, while retaining the ability to clean up when the instrument input is reduced.

Class Fuzz

Legendary pedal-whiz Roger Mayer used to modify the Arbiter Fuzz Faces used by Jimi Hendrix (see Fuzz Age). Though retaining much of the tonal color of the Fuzz Face (or here, the Fuzz Age), the Class Fuzz produces a more focused midrange than either Fuzz Age and a tighter breakup—more like a distortion pedal than a fuzz. At lower Drive settings, it will clean up nicely when backing off the guitar volume.

Control Issues The range of sounds available though the manipulation of the guitar volume and tone controls when using a fuzz pedal remains woefully unexplored by novice players. There is a tendency to plug in, crank the drive level, and fuzz out. Fuzz stompboxes can reveal many more colors if you take the time to experiment with setting different drive levels on the pedal and varying the volume and tone settings on your instrument.

- At lower drive settings, you may find that backing off your volume lets you play fairly clean chords, riffs, and licks. The sound you get will be distinctly different from that of a slightly broken-up amp or any other type of pedal.
- At higher drive settings, backing off the guitar volume will produce more of an overdrive or a milder distortion tone, unlike any you'll get with an amp or a non-fuzz pedal.
- Many fuzz pedals lack a tone control, so try rolling your instrument's tone control down. Even extreme tone roll-off can provide interesting results, especially when using maximum fuzz drive.

Fuzz One

This stomp simulation is based on the Maestro Fuzz-Tone. The legend of the fuzz box's origins involves a Nashville engineer named Glenn Snoddy. Snoddy built a transistorized box to emulate the sound that session guitarist Grady Martin got by playing through a busted mixing board. This fuzz ended up on a hit country record, causing other guitarists to seek that sound. Snoddy is rumored to have given the design to someone at Gibson. The company issued it under its Maestro moniker, and the rest is history. The world was introduced to the sound of fuzz through the Rolling Stones song "Satisfaction." Soon, guitarists such as Robby Krieger from the Doors and Steve Howe from Yes were spreading the Maestro Fuzz gospel.

The Fuzz One puts out less volume and fuzz than the Fuzz Age, but it offers a more metallic clang for edgy rhythm parts and cutting solos. This is a fuzz that works well in front of either a clean or a dirty amp.

Octa-V

Modeled after the Roger Mayer Octavia, an effect first used on "Purple Haze" and "Fire," the Octa-V mimics the original's frequency-doubling circuitry that produces a second note an octave above the fundamental note. It is a tone that can be heard on such tracks as "One Rainy Wish," "Little Miss Lover," "Little Wing," and "Machine Gun."

This is the most expressive of the fuzz voices, but it can be complicated to master. Here are some essential things to know when dealing with octave fuzz:

- The octave effect is most noticeable on the top E and B strings from the 7th fret up. In lower positions what you get is a metallic, almost ring modulator effect, which is cool in its own way.
- The effect on the Octa-V is more noticeable when you use less drive.
- Using the neck pickup and rolling back the tone will enhance the octave effect.
- It is more or less prominent through different amp and cabinet setups. It's much more pronounced through a Marshall with vintage 4×12 25-watt speakers than through a Fender 1×12.
- The pedal reacts very differently when placed before or after other fuzz, overdrive, or distortion pedals.
- Rolling off the volume while set on the treble pickup with the tone all the way on can create a creditable electric sitar sound.

Take the time to experiment with this pedal, and you'll find a wealth of unusual effects at your disposal.

RightFuzz

This effect is modeled after one of the earliest stomp pedals: the Mosrite Fuzzrite. Designed in 1966 by Eddie Sanner, the first 250 models used germanium transistors, but not being temperature stable, they didn't sound good on a cold floor, so Mosrite switched to silicon transistors.

"In-A-Gadda-Da-Vida" anyone? The riff on the Iron Butterfly hit was played through one of these. It offers almost a waveshaper sound with a bit of a frequency envelope that responds to attack, especially when using the bridge pickup. If you are using single-coils, this fuzz will benefit from a Diode Overdrive in front of it with the drive all the way down—just a little boost.

XS Fuzz

Post-1968, Jimi Hendrix began using the Roger Mayer Axis Fuzz, named after his album *Axis: Bold as Love*. The pedal is featured on the *Band of Gypsies* record. The Axis uses a more complex circuit configuration than the Fuzz Face.

Though Roger Mayer says that the Axis Fuzz produces more output level and sustain than the Classic Fuzz, the AmpliTube XS version is about the same in the sustain department (plenty) and, if anything, offers a little less gain than the Classic. When flat out, it offers more lows and highs, while the Classic leans to the smooth mids.

What the XS Fuzz provides in abundance is that Hendrix rhythm sound that you've tried desperately to get from various pickups, guitars, and amps but never quite matched. Simply set the XS Drive between noon and 2 o'clock, lower the volume on your guitar's neck pickup, and there it is.

Pitch

Sometimes guitarists record multiple lead or riff lines, whether octaves or harmonized, and then need to reproduce that effect live—without the aid of a second or third guitarist. This led the pocket-protector crew to come up with pedals that add octaves and harmony lines to a single riff or a solo played live.

In the studio, you don't necessarily need these devices to create multiple parts—you can just overdub them—but they have taken on a life of their own. Jeff Beck could have overdubbed the same part an octave lower for "Come Dancing" on *Wired*, but it would not have sounded the same as the pedal he did use. And though Trevor Rabin was no doubt capable of harmonizing his solo on "Owner of a Lonely Heart" through overdubbing, it would have sounded quite different than his use of a harmonizer.

Even with the high level of processing power available today, effects are unable to add identical-sounding harmony or octave notes to your originally played one. If you want the notes to match tone exactly, overdub them. But if you want to avail yourself of the character-laden tones of pitch-shifting effects, AmpliTube offers a number of them.

Caution: Although it is always important to have sufficient input level into AmpliTube for maximum sound quality, it is crucial to the proper performance of the pitch effects. If you're not getting a clear interval with any of the pitch stomps or rack units, check that your guitar volume is all the way up and that the input level on your interface is at the maximum that it can be without overloading.

Octav

AmpliTube's Octav stomp emulates the workhorse Boss OC-2 octave pedal that allows you to blend in a second signal one or two octaves below your original. As with the Boss, Octav gives you blend controls for one octave below, two octaves below, and the direct signal.

The Boss pedal, introduced in 1982, is not a pitch shifter in the digital sense, but rather it repeats the fundamental while emphasizing the harmonics of the octave below and two octaves below the original signal.

- This device allows you to fake bass parts by turning down the Direct Level knob to remove the original signal, add an octave or two below to emulate synthesizers, or just fatten the sound. Bassists can use it to add lower parts (if you don't have a five-string bass) or to keep the low end going when playing up the neck.
- A compressor in front of the Octav will help tracking by keeping the input signal even.
- The Octav wants to see a pure signal, so place distortions after it.
- Adding an envelope filter after the distortion can create a fat Moog bass-style sound (see Figure 8.10). Just be sure to remove the compressor and keep the distortion pedal gain levels on the low side so the envelope filter has plenty of dynamics to repond to.

Figure 8.10 This setup can create a creditable Moog bass sound.

Pitch Shifter

With the advances in digital processing that arrived in the '80s, pitch-shifting effects could now do more than add octaves. Any interval above or below was fair game, as well as the fractions between quartertones. Although at first available only in expensive rack units, such as the Eventide Harmonizer, they soon made their way into affordable pedals.

The AmpliTube Pitch Shifter stomp lets you choose any single interval from two octaves up to the same below your original note with the Coarse knob and blend in the amount with the Level control. The Fine control lets you adjust the pitch in cents, from 50 above to 50 below. Keep in mind that the values of 1 through 24 and −1 through −24 reflect half steps, not intervals—in other words, 7 means a fifth above, or 7 half steps.

- Though this device will not sound particularly good when strumming chords, unlike the Octav, you can play arpeggios that will track quite well, even if some strings ring over. This makes it great for emulating a 12-string guitar. I recommend a setting of an octave below and playing on the top three strings. This will sound like the lower end of a 12-string; tuning the Pitch Shifter an octave above reveals too much aliasing.
- Sometimes you might want the chime of a chorus but not the waver of its LFO. This is where the Fine tuner comes in. By shifting the pitch just a few cents and blending it with the original signal, you can get a subtler level of shimmer than with a chorus, but one that adds depth and width to a part without muddying it as much as a chorus might. *Detuning*, as this is called, can also be used to widen your guitar's stereo spectrum, but this involves stereo routing and panning, which we'll cover in Chapter 9, "Routing."
- You probably won't use the Fine tune when you are shifting full intervals via the Coarse tuning, unless you're seeking some detuned sonic mayhem.

Harmonizer

A standard pitch shifter maintains an unchanging interval relationship with the original note: No matter what note you play or whatever key you are in, the shift will be exactly what you set it at—in other words, 7 semitones, a perfect fifth. So, for example, if you were to then play in the key of E major, when you played the Major Seventh, or D#, the pitch shifter would play a perfect fifth—an A#, which is not in the key of E major.

A harmonizer, on the other hand, allows you to select the key in which you wish to play, as well as the interval, and then it maintains the proper relationships between the note you are playing and the harmonized note(s).

The AmpliTube Harmonizer is a three-voice digital harmonizer. It can create three separate harmony voices, each with its own interval value based on a musical scale. Once you select the scale to use, the voice number, and the desired pitch, the harmonized notes will be according to the selected scale.

In the aforementioned example, if you are using the AmpliTube Harmonizer with Voice 1 set for a fifth, and you have set the key for E major, when you play D#, the Harmonizer will play an A natural, which is in the key of E major.

Wharmonator

The invention of the DigiTech Whammy pedal in 1991 introduced a whole new sonic vocabulary into the lexicon of guitar expression. Now the intervals of a pitch shifter could be easily manipulated in real time. Soon, guitarists lacking vibrato bridges were dive-bombing notes and chords by setting the pedal to drop the pitch an octave or two when the expression treadle was pressed. Others were swooping up octaves or performing physically impossible intervallic leaps. A few used it as a "smart" harmonizer, shifting the interval as they played through a solo to match the key.

The Wharmonator does a bang-up job of emulating this versatile modern classic. As with the original, there is no Blend knob. On the Bend setting, it removes the original signal and lets you "bend" the note up or down one or two octaves, as well as down a second for whammy bar-style dives. In Harm[ony] mode, it adds in the original and lets you shift between different intervals in the heel and toe positions.

In addition to the effects described above, many others can add spice to your recordings. Here are a few:

- In Bend mode, you're not required to sweep the pedal. You can set it and leave it at an octave down for six-string bass sounds, while a fourth or fifth down will yield baritone guitar tones.
- In the same mode, down a fifth, you can swell notes in with a volume control placed after the Wharmonator to create cello parts. Remember to use a volume pedal after the effect, and not your guitar's volume knob, as the Wharmonator wants to see a steady input signal.
- Also in Bend mode, leaving it set at one or two octaves up, rolling off the tone on your guitar and swelling a volume pedal after the Wharmonator can create flute-like effects.

Modulation

Modulation is a term used when a pitch or volume effect changes over time. This change is usually brought about by an LFO (see the "LFOs" sidebar earlier in this chapter) modulating the effect at an intensity or rate determined by one of the unit's controls. The LFO will be modulating different parameters in different effects, but overall modulation effects such as tremolo, chorus, and flanging can add interest to static parts, fatten up thin ones, and contribute a wide range of colors to your recorded or live project.

Opto Tremolo

Tremolo refers to a controlled, rhythmic change of amplitude or volume, similar to rapidly moving either a volume knob or a volume pedal. This is called *modulating* the amplitude, which is why we deal with tremolo in the modulation section rather than the volume chapter—also because this is where you'll find it in AmpliTube. One method of modulation employing a pulsing light source shining on a photo resistor is used by the optical tremolo in Fender amps. The tremolo stomp in AmpliTube emulates the optical tremolo in a Fender Super Reverb.

- For more extreme modulated volume, you'll want to use the Step Slicer, but the Opt Tremolo is perfect for swamp, spaghetti western, and surf sounds.
- Combined with some reverb, it helps add an air of mystery to any music.
- Don't be afraid to use it on bass and baritone guitar parts.
- The BPM feature will sync the Opto Tremolo to programmed beats, but for more authentic roots stuff you may not want to sync it.
- Multiple tracks of synched tremolo parts, set for different beat divisions, can build interesting polyrhythms to lay over a straight beat.

Chorus

The AmpliTube Chorus stomp provides a rich, analog-sounding chorus when the Level is at noon, the Rate is at 10 o'clock or below, and the Depth is between 9 and 3 o'clock. More extreme settings can sound out of tune, but if you crank the Depth and Level up fully, while setting the synched Rate to 1/4, 1/8, or 1/16, you'll get a slightly dissonant, rhythmic warble that can be cool in some contexts.

Chorus Chorus effects are an attempt to reproduce the sound of a double-tracked guitar part. A chorus effect emulates this sound by splitting the signal into a dry one and one that uses an LFO to continually shift the pitch a few cents above and below the dry one by changing a short delay time. (You may have noticed with some delay pedals that if you turn the delay amount while a signal is passing through, the pitch of the signal changes.) The Depth control determines how far above and below, while the Rate control sets how quickly this shift occurs.

Exercise 5

1. Record a simple arpeggiated part onto Track 1 of AmpliTube's Recorder.
2. Retune the guitar anywhere from 10 to 35 cents lower with the tuner.
3. Now overdub the same part along with the first one. That is the sound of a Chorus.

Chorus-1

When the gray Boss CE-1 Chorus Ensemble (see Figure 8.11) was released in 1976, it was the first chorus effect available in pedal form. It was essentially the chorus/vibrato section of the JC-120 amplifier packed into a stompbox. It was an instant hit, adopted by such six-string luminaries as Joe Walsh and Lee Ritenour.

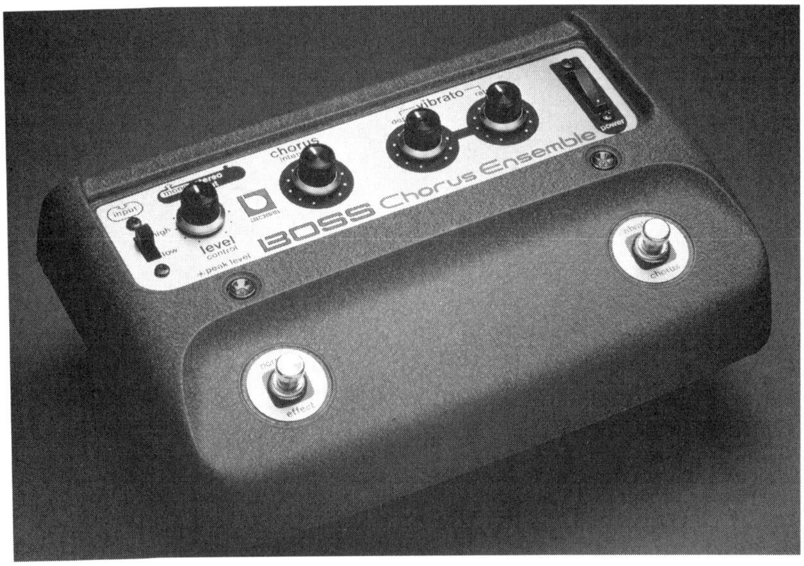

Figure 8.11 The first widely used chorus pedal: the Boss CE-1 Chorus Ensemble.

The Chorus-1 offers the original's switchability between chorus and vibrato. Though no rate control is available for the chorus, the Depth knob also increases the rate a bit as you turn it up. The Vibrato section has an independent Rate knob. The adjustment knobs on both sections remain musical through their full range, never producing any seasick detuning effects.

The Level control is a holdover from the original. This is not a blend between dry and wet signal, as it is on some chorus pedals. This is a true level control, increasing the overall output as you turn it up. Its purpose is to overcome the psychoacoustic effect of the chorus sounding quieter than the bypassed signal—when modulation effects are engaged, they can often make the ear hear the sound as quieter, even if it is measurably identical. Some effects hardwire in a slight boost as you turn them on. This stomp allows you to set the Level to match what your ear hears. Just be careful; there's enough gain to cause unwanted distortion at your amp's input stage.

Electric Flanger

In 1976, Electro-Harmonix brought out the first flanger pedal, the Electric Mistress, modeled here in the Electric Flanger. The Rate controls the speed of the LFO, while the Range is the E-H term for depth. The Color knob controls the amount of feedback—more feedback creates a more metallic sound. The Manual switch stops the LFO, letting you use the effect for static resonance and filter matrix effects. In this mode, you adjust the timbre of the resonance and filters by setting the Range knob, and you set the intensity of the resonances with the Color knob. This is a great setting for Robert Fripp–like metallic rhythm parts, a la late King Crimson or *Low*-era David Bowie.

- Don't be afraid to experiment with extreme settings. The Electric Mistress is chock full of character and will yield a wide range of tones from pretty to scary.

- Much of what sounds like chorus on the early Police records is an Electric Mistress. Try the settings in Figure 8.12 for an Andy Summers sound.

Figure 8.12 This configuration should bring the Police running.

Flanging Flanging began in the experimental '60s, when recording engineers would play the same signal into two tape machines while holding their thumbs on the *flange*, or the outer edge of the tape reel. This delayed the recording on that reel, and when the reels were played back together, the short delay would create a comb-filtering effect. A psychedelic *whoosh* resulted that can be heard on Hendrix and early Small Faces records.

In a flanging pedal or rack effect, the signal is split in two: a dry signal and a second signal that is delayed by a number of delays in the 5–10ms (millisecond) range. The delay is modulated by our old friend the LFO and fed back on itself, creating the signature whooshing.

Analog Flanger

Boss put out an early flanger called the BF-1 in a large gray housing, and in 1980 the company replaced it with its new compact pedal size, the box we have all come to know and love—Boss called the new model the BF-2. It had the same controls and quickly became one of the most widely used flangers around, popping up on the pedal boards of artists such as Neal Schon, Jake Lee, Yngwie Malmsteen, and Steve Stevens.

AmpliTube offers no fewer than four flanger stomps, and with this one we start to see the propensity of manufacturers to label the same controls by different names. Here, what the Electric Mistress calls Range is called the more accurate Depth. Rate is as advertised, while feedback is called Res, for resonance. Manual sets the timbre of the flanging effect, and if you turn the Depth knob all the way off, you can get the same static comb-filter effects as with the Manual switch setting of the Electric Flanger.

Though both the BF-2 and the Electric Mistress are analog flangers, using bucket brigade delay chips, the BF-2 is a darker, warmer whirl, while the Electric Mistress offers a brighter, more sparkling swirl.

Metal Flanger

The original MXR M-117R Flanger was used on Van Halen's "Unchained" and "And the Cradle Will Rock," thus becoming the go-to flanger for metal-heads worldwide. Metal Flanger models this iconic unit, offering the same controls. Here Depth is called Width, for some reason known only to the original designers at MXR. Regen (for regeneration) is another way of saying Feedback, and Rate becomes Speed. As with the Analog Flanger, turning the Width control all the way down lets you create static effects.

As befitting a metal flanger, this unit had an edgier sound than the Mistress or the BF-2. If you are cutting a metal track and need flanging, you might want to start here.

Flanger

Don't let the generic name of the Flanger fool you. It may not do static comb-filter effects, but it offers rich flanging, and only it and the Electric Flanger let you sync your flanging sweeps to the beat. If you need that last feature and something a little fuller sounding than the Electric Flanger, you've come to the right place.

Small Phazer

One of the most popular phasers of all time is the Electro-Harmonix Small Stone (see Figure 8.13), modeled here. It is a four-stage phaser, designed in 1974 by David Cockerell, who was hired by Electro-Harmonix from his former employer, EMS. It uses operational transconductance amplifiers (OTA) and has one additional stage for feedback, which can be activated using the Color switch. An OTA-based phaser produces strong third-harmonic distortion, which is part of its warm analog sound.

Like the original Small Stone, the Small Phazer offers the Color switch and a Rate knob, but unlike the original, a switch allows you to sync this rate to your project's BPM.

When artists such as Radiohead's Jonny Greenwood or Television's Richard Lloyd want some swirl, they turn to this pedal.

A cool way to use phasing when soloing is to place it after the delay and route the delay in parallel so the original signal goes through undelayed and unphased, while the parallel signal goes through the delay first and is then phased. We will get into routing in depth in a later chapter but here is a preview:

Figure 8.13 One of the earliest and still most popular phasers.

Exercise 6

1. Set AmpliTube for Routing Path 5 and place a Distortion pedal in Stomp Path A.
2. Save the settings of the Distortion as a preset by right-clicking on the pedal. Call it Phase Dist.
3. Place Distortion in Stomp Path B and call up the Phase Dist preset.
4. Add the Delay stomp to Stomp B, setting the Delay for 1/4t and the Feedback at about 10 o'clock. Set the Level to entirely wet.
5. Place a phaser in Stomp B after the Delay. Set the phaser to a slowish sweep—I'm using Small Phazer here, but any phaser will do.
6. Now set the volume of Stomp Path B a little lower than Path A in the Selected Module section—this is where you adjust your wet/dry amount to taste. Your Stomp Path B setup should look like Figure 8.14.
7. Now when you solo, your original note will come through distorted but dry, while a phased delay follows a quarter-note triplet behind.

Phasing Like the chorus and the flanger, the phaser or phase-shifter splits the sound in two, shifting one of the signals out of phase with the original. When the peaks of one signal line up with the valleys of the other, the sound all but disappears. The second signal is treated with all-pass filters, each frequency with a different short delay (10–30ms). An LFO alters the filter frequency ranges over time, and this continually shifting phase creates the swooshing sound that we call *phasing*. Most phasers employ assorted peaks and notches taking place at different points in the timeline, called *stages*, ensuring that the signal is never phase-canceled out of existence.

Figure 8.14 This setup will produce a delayed phased signal.

Phaze Nine

The Phaze Nine stomp is a model of the Small Stone's strongest rival, the MXR Phase 90, which appeared on the market at roughly the same time. It uses more expensive field effect transistors (FETs) to create the phasing effect, giving the E-H Small Stone an edge in the market. Because FET-based phasers have a pronounced second harmonic distortion, the Phase 90—and thus the Phaze Nine—has a distinctly different tone than the Small Stone or Small Phazer. It sole knob, Speed, makes it the essence of simplicity to use.

- Phasing in not just for psychedelia and metal jet-plane whoosh, it adds a cool envelope to funk chording and picking as well. The Phaze Nine has a warm sound best for funk in smaller combos. With larger bands or denser sonic situations, you may find that the Small Phazer cuts through better.
- In the late '70s, phasing started showing up on country records as guitarists such as Waylon Jennings and James Burton took up the effect.

Phazer 10

For the Phase 100 pedal, MXR added a notched Intensity control that depicted different wave shapes. In general, this pedal has more of a chorused sound than either the Phase 90 or the Small Stone. The second Intensity setting somewhat resembles the Small Stone with the color engaged, only more vocal-sounding, almost like a talk box. With the Phazer 10, you can get a wide range of phasing effects from mild to wild.

Uni-V

Originally designed to emulate the slow speed of a Leslie, the Uni-Vibe (see Figure 8.15) is familiar to guitarists largely through its use by Jimi Hendrix on "The Wind Cried Mary" and Robin Trower on "Bridge of Sighs." It is technically a type of phaser that uses transistors, four light bulbs, light cells, and an LFO to do the shifting. Reflective light shields ensure the correct diffusion of light from the bulbs. This method creates a phasing effect that is more luxuriant than most phasers.

Figure 8.15 The Dunlop reissue of the Uni-Vibe has the same controls as the Uni-V.

Stevie Ray Vaughan's revival of the Hendrix sound and Robin Trower's new recordings and touring have kept the 'Vibe in the public eye since its introduction in the '60s. A slow Uni-Vibe can add a dark mystery to a clean sound, while faster speeds lend a psychedelic swirl. A Uni-Vibe sound tends to work better before the breakup of an amp or pedal—the grit helps bring out its lushness. The Uni-Vibe is preferable to chorus or flanging for fuzz lead work, as it adds character without obscuring the lines.

As with the CE-1 Chorus' Level control, the Uni-V's Volume is a true volume control, not a blend, and serves the same purpose of counteracting any psychoacoustic dip from kicking in the effect. Intensity and Rate are self-explanatory, but CHO/VIB may be less so. The original Uni-Vibe was touted as emulating the Chorale setting on an organ, so CHO does not stand for chorus, but rather chorale. This is the Hendrix/Trower sound we all know. The VIB setting is more like a Magnatone amp's vibrato, familiar from Lonnie Mack's hit instrumental version of Chuck Berry's "Memphis."

The Uni-V does a terrific job of conjuring up the original Uni-Vibe. The only issue is that it's hard to play anything through it without sounding like the guitar legends who used it. Feel free to accept the challenge, keeping in mind that the vibrato setting sounds great as well and brings with it much less historical baggage.

Rotary Cabinet 1 and 2

Why is a cabinet in the effects section? Well, it's unlikely that, as a guitar player, you'll have a rotary cabinet as your primary speaker cabinet. More likely, you'll use it as a special effect, like the one heard on recordings such as Cream's "Badge," the Eagles' "Hotel California," the Grateful Dead's "Casey Jones," and many others.

A number of pedals emulate this sound, but before they existed guitarists would wire up a switcher that would let them go from their normal sound to a Hammond organ's Leslie cabinet.

The Leslie sound is created with a pair of treble speaker horns pointing in opposite directions and a bass speaker pointing down. Only a single treble horn is functional; the other is for balance as the pair rotates. A baffle surrounding the bass speaker also rotates, only letting sound out through one hole. The operating treble horn moving toward and away from your ear creates a Doppler effect, like the changing pitch of a train's horn as it passes. The baffle letting the bass driver sound pass through only at certain intervals creates a kind of tremolo. The combination of these two effects is what we recognize as a "Leslie" sound.

AmpliTube 3's VRM (volumetric response modeling) technology helps create a swirling, phasey, almost underwater-type sound remarkably like the original. The Rotary Cabinet 1 contains an original Alnico speaker for a warmer, softer sound, while 2 sports a modern replacement woofer for more presence and clarity.

- A Width control lets you determine how stereo or mono the sound is. Full stereo makes for a cool psychedelic motion across the soundstage spectrum, while mono lets you focus the sound more in a complex mix.
- The Balance knob lets you choose more of the woofer or more of the tweeter; it's a great way to sit the sound in the mix before you have to touch any EQ.
- The Speed control chooses Fast, Slow, or Brake. As with the original Leslie, the speed ramps up gradually to Fast from Slow and slowly retards to Brake from Fast before you can switch it to Slow. If you stay in Brake mode, the speakers are not moving but will lend a distinctive coloration to the sound.
- You can control the speed with a MIDI pedal for "ramping" effects, much like an organist.
- The tweeter and woofer are pre-miked; you can choose more tweeter or woofer but not the type of mic. The room mics are in effect (in Hi and Mid quality modes) and can add depth to the sound.

Because with AmpliTube you don't have to rewire the inputs to accept guitar or carry around the 80-pound cabinet, you may find yourself returning to this lush effect again and again.

The Effects Chain One thing guitarists who are new to the world of effects always ask is, "In what order do the effects go?" AmpliTube allows you to place up to 12 effects, in any order that you like, in front of the amp and up to 8 effects, also in any order, after the amp. The basic rule is that whatever sounds right to you is fine; feel free to experiment. AmpliTube lets you easily drag and drop the effects for comparison purposes.

Here are a few tips that might guide you in your choice of which effect goes where:

- Hardware fuzz boxes of the vintage germanium transistor variety must be placed first in the chain, with the guitar going directly in, as they interact with the changes in signal from the volume level of the guitar. This interaction is interrupted if another pedal is placed in between. This is not as pronounced in AmpliTube's software versions, but to be safe, put them first.
- Auto or envelope filters come next in the chain. They want to respond to the full dynamic range of the guitar attack. Any compression by compressors, overdrives, or distortions will dilute their effect.
- If you're not using a germanium fuzz or envelope filter, stompbox compressors usually come first in the chain, as they will add noise to anything that comes before them. (Due to its software nature, AmpliTube doesn't react like this, but keep it in mind for hardware.) Rack compressors can be used post-amp and in mixing, as they are higher quality and used more subtly, thus adding less noise.
- I recommend that you start by putting a Whammy pedal second. DigiTech recommends that you place it before any distortion so that it can track a clean signal. I've had interesting results both before and after.
- Next you might want your wah-wah pedal. Amps and pedals break up differently depending on the frequency of the signal pushing them. Many classic rock tones have been created with the wah driving amps and distortion pedals with the differing frequencies the wah creates. If you place the wah after a distortion pedal, the pedal always sees the same frequency at its input and will break up in the same way, regardless of the frequency coming out of the wah pedal.
- On the other hand, this may be what you want. Placing the wah after distortion pedals makes it act more like a synthesizer filter—changing the frequency of the distorted signal. See which you like better.
- Now we come to the distortion pedals. Many players use more than one, either individually or driving each other. No rules—just experiment.
- Normally, modulation pedals would come post-distortion, as overdrives and distortions like to see an unsullied signal. Too, the compression and extra harmonics of distortion come alive when run through a chorus or flanger.

- Delay often comes toward the end, once again to let the overdrives receive a pure signal. In the early days, tape delays were often put in front of distorted amps; so if you're going for that sound, by all means try it. Another exception is if you're using a chorus to split your sound into stereo—then you would place the delay before the chorus. For the adventurous, there are also some interesting effects to be had by filtering and modulating the delays with filters, flangers, and choruses.
- Reverb almost always comes last, save for more experimental musicians, who have been known to place distortion and filters afterward. Placing distortion after the reverb is not an option in the basic AmpliTube, but it can be done if you have AmpliTube Fender, where a spring reverb appears in the Stomp section.

Rack Effects

AmpliTube 3 offers a whole other set of effects available in the Rack section of the routing chain. In a physical studio, these are the effects you use after you send the signal from the microphone recording your amp to the mixing board. They are often mounted in racks and patched into inserts in the mixing board. Your guitar sound can be routed to them during either recording or mixing.

When recording with AmpliTube, you can use some of these effects as an integral part of your sound and others to shape the final sound during the mix. Some of these effects are similar or identical to available stompboxes; nevertheless, as you learned from "The Effects Chain" sidebar, the sound can be quite different depending on where the effect occurs along the routing path. Tap Delay, Rezo, Step Filter, Analog Chorus, Rotary Speaker, Harmonator, Pitch Shifter, Step Slicer, and Swell more or less duplicate effects from the Stomp section. The difference is that here they have no influence on how the front end of the amp reacts.

Here is an exercise that will demonstrate the difference between placing an effect before and after the amplifier. (This is not a particularly great sound, but I've used it because the difference is obvious.)

Exercise 7

1. Place a Rezo effect in the Stomp A section with the default settings except for the mix, which should be turned up all the way (see Figure 8.16).
2. Place a Rezo effect in the Rack A section with the default settings (the same as the stomp) except for the mix, which should be turned up all the way.
3. Choose a British OR amp and set the controls as in Figure 8.17.

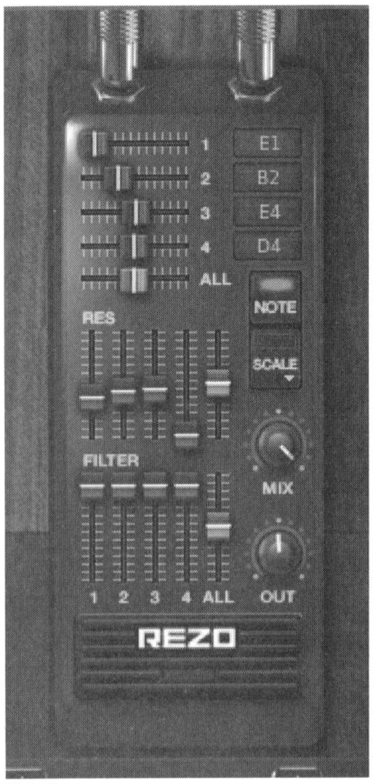

Figure 8.16 The Rezo settings of both the stomp and the rack should look like this.

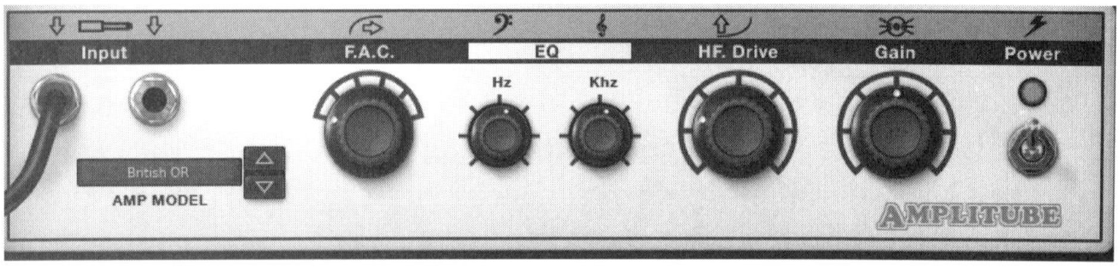

Figure 8.17 Set the amp controls thus.

4. Shut off the rack Rez and play an open G chord. Note how certain frequencies are emphasized by the reaction of the amp and cabinet to the resonance of the stomp effect.

5. Now shut off the stomp Rez, turn on the rack version, and play the same chord. Note how the sound is more evenly distributed across the frequency spectrum. This is because the effect is modifying the sound coming out of the amp, rather than affecting the sound driving it.

There are a few differences in the stomp and rack versions of the aforementioned effects:

- The Pitch Shifter and Harmonator offer stereo effects in the rack version but not in the stomp version. An interesting quirk of the rack pitch shifter is that the default setting—with no pitch shifting and the stereo switch set to Split—instantly sets up a cool, hard-panned left and right, doubled effect. Also, if you use the amp for distortion, you'll definitely want to use the rack Harmonator rather than the stomp version for best tracking.

- The Rotary effect does not offer the stereo width adjustment in the rack mode. It does, however, provide a continually adjustable speed control. You lose the vintage ramping possibilities unless you hook up a MIDI expression pedal to the speed control and do your own ramping. The Balance control is here, as well as a new Drive control to adjust the amount of Leslie grit. An advantage of the rack version is that you can use your amp's speaker as well as the Rotary one—an effect mix control lets you blend the two.
- Like the Chorus-1 stomp, the rack Analog chorus will do chorus or vibrato. It offers a stereo option not available with the stomp but no BPM sync.

The Rack section also adds some new effects to the mix.

Digital Delay

This is a great effect for spreading your sound across a mix without having to set up any kind of stereo routing path.

- ST helps maintain the stereo spread when you are using a stereo effect in front of the Digital Delay.
- LR, or Left Right, ping-pongs the delayed signal back and forth between the left and right sides of the stereo spectrum while the dry signal comes up the middle.
- MONO places the delayed signal behind the dry signal in the center.
- LCR, or Left Center Right, sends the delays left, then center, and then right. In other words, if your BPM was set to quarter notes, and you hit a staccato note, the dry signal would play down the middle on Beat 1 (if that's where it was panned), the first delay would appear on Beat 2 on the left side, the second delay would be down the center on Beat 3, and the third delay would be on Beat 4 on the right. Even if you have the feedback set to one repeat, you will hear the left, center, and right delays. This a terrific setting for lush, swelling pads, whether you use a volume pedal, the Swell effect, or the Volume knob on your guitar. Be sure to use the low-pass filter to roll off some highs.
- DOUBL[e] sends the delay simultaneously left and right. Using a short delay of 5 to 25ms will fatten out your solos without adding any potentially muddying pitch shift. You'll have to switch off the BPM setting to get delays this short at any but the fastest synched tempos.

Digital Reverb

Unless you have the Fender package, AmpliTube doesn't offer any reverbs in the Stomp section. Most of the amplifiers have a spring reverb available (even if the original amp didn't), but if you want lush digital reverb effects, the Rack section is where you'll get them. This makes sense, as placing reverb in front of all but the cleanest amp is less than ideal except for special effects. The rack Digital Reverb in AmpliTube offers gorgeous ambience for pads and moody parts.

- When recording, you should avoid committing big reverbs to the track unless you are planning to build the rest of the track's tone around them.
- If you have enough CPU to leave the AmpliTube plug-in running all the way through the mix—no problem; you can adjust the reverb later, when mixing.
- If you must commit the sound before the mix, it is better to leave the AmpliTube reverb off and apply studio reverb later so it can be adjusted to the sound of the other instruments.

The AmpliTube rack reverb sounds so good you might want to put it in an effects return in your DAW and use it on other instruments as well.

Graphic EQ

We discussed graphic EQs in the stomp section. The rack Graphic EQ is like the stomp versions on steroids: Rather than 7 or 10 bands of equalization, it has 31! This lets you focus in tighter on specific frequencies that you might want to adjust. This is where you can shape the tone of your overall guitar sound without affecting how the amp responds.

- As I recommended with the stomp EQs, any adjustments made here should be small. If you find yourself looking at any major curves, increases, or cuts in frequency, you may want to reconsider your choice of amp or cabinet and/or your microphone choice and placement. Fixing those things will result in a more natural sound than radical EQing.
- If you do perform any radical cuts or boosts, be sure to readjust the volume with the EQ's Level control.
- We went through the stomp EQ frequency sliders to learn which parts of the sound reside at which frequencies. Doing the same thing here will fine-tune your ears and help immeasurably when using the next tool in the rack: the Parametric EQ.

Parametric EQ

A parametric EQ lets you home in even more precisely on frequencies that you might want to boost or tame. The rack Parametric EQ offers three basic ranges: Bass, Mid, and Treble. Within those ranges, the Freq[ency] control lets you choose a very specific frequency to boost or cut (it will read out below on the Input/Output bar)—for example, 327.8 Hz. The Q control lets you choose the bandwidth of the bass parametric EQ, from 0.1 (very wide) to 8.0 (very narrow)—that is to say, how much of the frequency range on either side of this amount will be affected. The Gain knob chooses the dB of the boost or cut you want.

Exercise 8

Here is how you use the Parametric EQ:

1. Figure out the general range of the frequency you want to work on.
2. Then set the Q at about noon. Set the gain at around 10 dB.
3. Now sweep through the frequency ranges while the part is playing. (If you haven't recorded it yet, you may need another pair of hands to do the sweeping while you play.)
4. When you hear the frequency under consideration boosted in this exaggerated manner, you may then either boost it appropriately or cut it to taste, depending on what the sound needs.

Tube Compressor

This model emulates classic tube compressors of the LA-2A or Fairchild 670 variety. Those compressors are used as much for color as for compression, and their effect is subtle. I recommend that first you go back to the "Compressor" section at the beginning of this chapter to refresh your memory on the meanings of some of the terms. Then try out this effect after you've gotten the sound the way you want it in every other respect.

- When adding compression at this stage, make sure you adjust the output level so that the volume is identical with the compressor on and off. That way you'll hear whether the compression is actually adding something to the sound or just making it louder.
- On the other hand, making the signal louder and more present is one of the things that this type of compression does well, so feel free to use it for that purpose if it suits.

Digital Chorus

Aside from working better for chorus sounds with a distorted amp, the Digital Chorus offers controls that are unavailable from the stomp pedals. Thanks to a choice of LFO waveforms, you can create chorus textures that range from subtle sine-wave sweeps to wild square-wave chops. The Mix control lets you blend in as much or as little of the created effect as you like.

Digital Flanger

As with the Digital Chorus, you get a choice of LFO waves here. Experimenting with sine-, triangle-, and square-wave based flange effects will produce the standard chime and whoosh, but also extreme pitch sweeps, resonating squeals, and synth-style filter effects.

Caution: The Rate, Depth, and Feedback controls are extremely interactive, so be careful when constructing sounds. Extreme resonances can harm speakers and ears.

Stereo Enhancer

This rack unit is for widening or narrowing the stereo field of your sound. It appears to widen the sound by lowering the volume of whatever is in the middle, thus emphasizing whatever is panned left and right. It narrows the sound by lowering the volume of whatever is in the left and right channels. I can see getting more use out of the narrowing function than the widening one. If you have a stereo preset that you love, but it's taking up too much space in the mix, this effect could be just the thing to slot it in.

Summary Questions

You can find the answers to the Summary Questions in the Appendix at the end of this book.

1. What is the main rule when using effects?
2. What is the definition of the term *dynamic range*?
3. How does a compressor work?
4. Why is it called a "graphic" equalizer?
5. What is the difference between a high-pass, a low-pass, and a band-pass filter?
6. What is the difference between an envelope, an LFO, and a step filter?
7. What is the difference between boost, overdrive, distortion, and fuzz?
8. Where is an Octa-V's upper octave most noticeable?
9. How does a pitch shifter differ from a harmonizer?
10. What is the difference between chorus, flanging, and phasing?
11. What is the difference between graphic and parametric EQ?

9 Routing

When high-budget, major-label records are made featuring guitar, it is not unusual for the parties involved (performer, producer, and engineer) to set up a number of amps with a splitting system that allows the performer to play a part through all of them simultaneously. The producer and/or mixing engineer will then have the option of blending these amp sounds together into one glorious tone. Sometimes they will mix this sonic cocktail right on the spot and commit the sound to a single track; other times they will save the decisions for the mix.

When recording with AmpliTube, you have a number of different ways to stack sounds in this manner. The Module/Rig Selector section lets you choose eight different signal paths employing up to two separate amplifiers and cabinets. All of this will get sent to one stereo track see (Figure 9.1).

If there are plenty of tracks available and enough CPU power, you can stack as many amps as there are in AmpliTube, each on its own track, playing your part into all of them at once (see Figure 9.2).

But if CPU or track availability is limited, these eight routing options will still allow you to easily stack two independent sounds, as well as provide a plethora of parallel processing options.

Let's go through AmpliTube's routing presets one by one and discuss the advantages and possibilities of each.

Series or Parallel In this section you will constantly hear the terms *series* and *parallel*. These terms are used in wiring circuits and pickups, in connecting speakers in cabinets, and in routing signal paths.

- Series means that the signal is going out of one source directly into another source.
- Parallel means that two signals are traveling alongside one another.

If you were to run a fuzz pedal and a delay pedal in series into a single amp, the guitar signal would travel from the instrument into the fuzz, then out of the fuzz into the delay, and then out of the delay and into the amplifier. Running the delay pedal in parallel means that after the signal comes out of the guitar, it is split in two: One part goes through the fuzz, and the other goes into the delay. After leaving the two pedals, the respective signals would be joined together again and enter the amplifier. The signal going through the delay would be clean, never having passed through the fuzz.

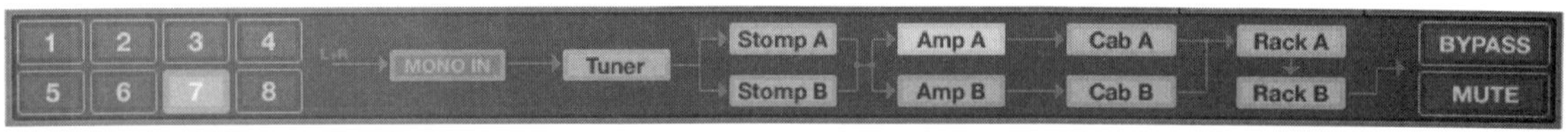

Figure 9.1 AmpliTube's Module/Rig Selector section.

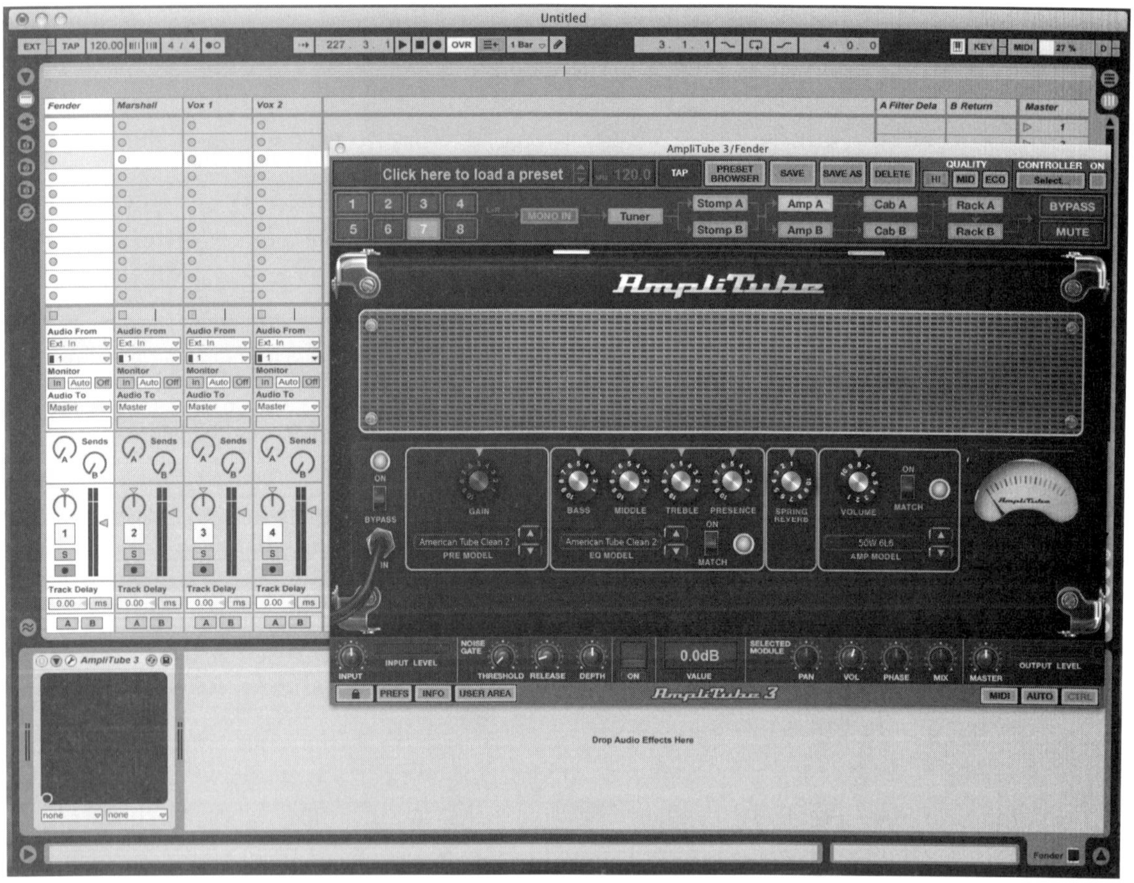

Figure 9.2 Here is a setup that lets me record four different amp tones at once to be blended later.

Preset 1

In this preset your stomp effects go through a single amp and cabinet and then into the rack effects. This is similar to how you would play live in mono, with rack effects applied at the board, or how you might quickly record a mono track in the studio. With this preset everything travels in series, one into the next. This is the routing with which many players probably are most familiar.

- This routing makes for a focused sound that is easily panned into a specific location in the stereo spectrum.
- This is a good preset if you're trying to create a specific vintage sound. You will want to use a single recognizable amp sound in mono—the way they did way back when.
- With this preset you can run up to 12 stomp and 8 rack effects, one into the other.

Preset 2

With this preset we enter the world of parallel routing. Here you have the choice of running a single instrument through two entirely separate rigs. If you are really pressed for time or tracks, in stereo input mode you can put two guitarists or a guitar and a bass onto one track. For our purposes, let's assume you are tracking alone.

This routing can get very complicated very quickly. Perhaps the best way to think about it is to pretend that you are two different guitarists, each with his or her own rig. Alternatively, you could think of it as if you were one guitarist overdubbing the exact same part but after switching rigs. Here are some tips for making the most out of this preset.

1. In the Selected Module section of the Input/Output bar, pan the two rig paths hard left and right in both the cabinet module and the rack module. For now we will pan the A cabinet and rack modules in the same direction and the B pair in the opposite direction. You can eventually create interesting effects with cross-fading, but right now we want to hear each rig on a single side.
2. Choose the Cab and Rack modules in Rig B and click on Mute at the end of the Module/Rig Selector section (see Figure 9.3). You should now hear only Rig A coming out of one speaker.

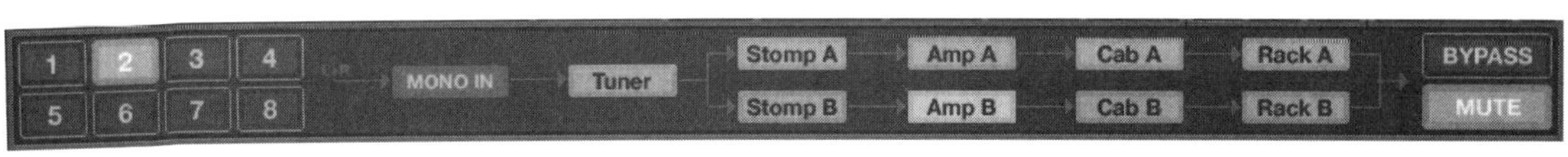

Figure 9.3 Clicking on any of Rig B's modules and then Mute will mute the rig while you work on A.

3. Now adjust all of your settings for Rig A: stomp selection and settings; amp and cabinet choice and settings; mic choice and positioning; and any rack effects for that rig, with their settings. Pay particular attention to input and output levels, making sure that you're not driving the output into digital distortion.

4. Mute the Rig A cabinet and rack selected module.
5. Unmute Rig B and repeat all the same adjustments.
6. Unmute Rig A. The balance between the two rigs may not be as you would like. *Do not turn up the quieter rig; turn down the louder one.* Turning up the quieter one will overload the Master output.
7. Now that you have the rigs sounding the way you want and relatively balanced, you may not want them panned hard right and left. You can pan them in any relation that you like; just keep in mind that you may need to further adjust the balance and tones once you reposition them.

There are many reasons to set up separate rigs like this. You might want to increase the power of a metal chord rhythm by having it played through two amps. You could use two matching Metal Lead V's, but try this exercise first.

Exercise 1

1. Set up a Metal Lead V in both Rigs A and B, pan A and B hard left and right, and play the part.
2. Now switch Rig B to American Lead Mark III and listen. Note how the panning becomes more distinguishable due to the difference in the way the amps sound.
3. Start scrolling down through the Lead amps. Listen to each one and note how it affects the panning and the overall sound. Feel free to experiment with Clean and Crunch rigs as well as with Rig B's volume. Placing it behind A will sound very different than keeping the two paths equal.
4. Also experiment with panning. A highly different and more present sound, such as British Tube Lead 1, can easily dominate Metal Lead V in the mix when its volume is equal and they are hard panned; but pan them closer and lower British Tube Lead 1's volume, and you will notice how it just adds presence to the Metal Lead V tone.

A common studio trick is to add definition to a crunch rhythm part by pairing a slightly overdriven amp with a clean one. It takes only a slight readjustment in the balance between the two, and the same pairing can add a little grit to a chiming clean sound.

When going through AmpliTube's stomp effects, you may have noticed that most of them lack any kind of blend control; when you turn on Chorus-1, it is all or nothing. Sure, you can often adjust the intensity of the effect, but barring the delay effects, the amount of dry signal versus wet signal remains constant no matter how you adjust the effect's controls. Preset 2 has an answer to this problem.

1. Set up a pair of parallel rigs in Routing Preset 2 that are identical save for a stomp effect in Rig B.
2. Presto—you can now adjust the amount of effect relative to the dry signal by raising or lowering the volume of Rig B. (You can also do this in presets 5, 6, 7, and 8.)

Preset 3

Dual-cabinet rigs go back to the early '60s, when the Fender Dual Showman was wired with a 4-ohm output so that you could use two speaker cabinets. The later '60s brought us the Marshall stack, with Jim Marshall angling one of his 4 × 12 cabinets because he thought it looked better when stacked on a straight cabinet and might be more stable.

For the most part, dual-cabinet rigs are a live phenomenon developed to throw a more dispersed sound at the audience. As we discussed in Chapter 6, most often only one speaker of a cabinet is miked up, meaning that kind of dispersion is audible only through distant room mics.

However, there are reasons to use more than one cabinet when recording.

- The aforementioned distant room recording will pick up the difference in sound between a single cabinet and two.
- Using cabinets with different speakers can create a similar effect to using multiple amps, even when miking only one speaker in the cabinet, provided the speakers are different enough. These effects include panning for stereo spread and filling in frequency ranges that each cabinet lacks.

Preset 3 allows you to use two cabinets with a single amp, each with its own miking setup. As we discussed in Chapter 7, "Miking," any time you use more than one microphone on a source, you must be wary of phase cancellation. Using two cabinets in AmpliTube is no exception. Here you have potentially eight microphones in play: two close and two room mics for each cabinet.

Here is one way to approach miking up two cabinets:

1. For simplicity's sake, I recommend that you start with the room mics off.
2. Set Microphone 2 on both cabinets to None, using only one mic for each cabinet.
3. Decide which frequencies each cabinet does best and match the cabinet with the microphone that ideally captures those frequencies.
4. Bypass Cab B and work on placing Cab A's mic in the optimum location; then bypass Cab A and do the same for it.

5. Enable both Cabs and listen to the sound. Try placing one of the mics out of phase. Does it sound better to you or worse? Try adjusting the Phase knob (see Figure 9.4) in the Selected Module section to various positive values. This will delay the signal from that cabinet and affect the phasing as well.

Figure 9.4 Setting the Phase knob to a positive value delays the signal from that cabinet.

6. Panning the cabinets will have an effect on the tone as well.

This doubles the painstaking process that every studio engineer goes through when miking up a cabinet. It is complicated and time consuming in AmpliTube as well, but considerably less so, for as we mentioned in Chapter 7, you don't have to keep running back and forth between the control booth and the cabinets.

One instance when you might want to use this preset instead of Preset 2's dual rigs is if you are using a bunch of effects in series into an amp with a sound you like, but you would like to create a stereo spread. In Preset 2 you would have to match all the stomp settings in both rigs, whereas here you have just one set of pedals and can spread the speakers left and right. The Phase control comes in handy here as well, because adding some delay to one of the cabinets enhances the stereo effect.

Preset 4

Preset 4 sends all stomp effects and rack effects into two amp rigs. This preset has many of the advantages of Preset 2, but, like Preset 3, it is preferable if you want to use one set of up to 12 effects into both amps. It also lets you use up to eight rack units in series, as opposed to Preset 2's parallel setup.

Preset 5

Two separate stomp and rack effects setups into one amp with two cabinet setups offers the advantages of two cabinets and parallel effects routing. Another effective way to use parallel routing is to create a complex delay and reverb setup that offers a subtle ambience in Rack A alongside a more extreme one in Rack B.

Exercise 2

- In Rack A, set up a delay and reverb with settings like you see in Figure 9.5. The eighth-note triplet setting is for the delay. This will give your guitar a little bit of immediate depth.

Figure 9.5 These settings in Rack A yield a subtle ambience.

- Set up Rack B like you see in Figure 9.6, with the delay at a quarter note and the mix all the way up. This way, the heavier reverb with longer decay will not kick in until a quarter note after the played note.

Figure 9.6 This setup brings in the intense reverb well behind the played note.

This is similar to the more complex settings that a studio reverb allows, with Rack A acting as the ambience on the predelay created by the digital delay in Rack B.

Preset 6

This preset may look very much like Preset 2, but it is not quite the same. In Preset 2, each parallel route—stomp, amp, and cab—goes through its own rack only; Route A goes through only Rack A.

In Preset 6, stomps, amps, and cabs from A and B go through both Rack A and Rack B. Because the racks are still in parallel, the effects in Rack A do not go through Rack B.

Preset 7

In Preset 7, two parallel stomp effects setups are sent to both separate amp rigs and then into all the rack effects in series. This is a great routing that combines sonic girth along with the flexibility of two separate amps and cabinets and parallel stomp routing while being simpler to set up than Preset 2.

One thing that makes Preset 2 complex is that if you want the same overdrive going into both amp routes, you need to set up the same stomp in both Stomp A and Stomp B and make sure that the settings are the same. Here a single overdrive inserted into Stomp A will go through both Amp A and Amp B.

Even though Preset 7 sends both Stomp A and Stomp B through both A and B amps, you still get the advantage of parallel effects routing. You can still place modulation and filter effects in Stomp B and adjust the amount of the effect in relation to the dry signal by adjusting the Selected Module volume.

Preset 8

This preset gives you the ultimate simplicity of a single-amp setup while retaining the advantages of a parallel-stomp arrangement.

Here is an exercise that will demonstrate how effects run in parallel can sound very different than the same run in series.

Exercise 3

1. Set up Stomp B to look like Figure 9.7. Use an American Vintage T amp, as it will handle the wide range of filter frequencies without breaking up.
2. Then set up Stomp A identically except without the Step Filter. Shut off the effects in Stomp A.
3. Play whole notes and listen to the effect of the LFO Filter running through the Step Filter in Stomp B.
4. Now turn off just the LFO Filter in Stomp B and turn on both effects in Stomp A.

Figure 9.7 The Stomp B setup should look like this.

5. Play the same whole notes, and you will hear how different the same effects sound when routed in parallel.

Summary Questions

You can find the answers to the Summary Questions in the Appendix at the end of this book.

1. What is the difference between series and parallel routing?
2. Name two reasons to use routing Preset 1.
3. Which routing presets allow parallel effects routing?
4. If you just want to have two cabinets, which preset would you use?
5. Which preset lets you use two separate amp and cabinet rigs in parallel, but still use all the effects in series?
6. If you wish to run parallel stomps with two separate rigs and series racks, which preset would you choose?

10 Tone Tips: Sound Like the Stars

To a large extent, the joy of playing music is linked to the thrill of discovery. By experimenting with different amps, effects, and miking options, we come upon sounds that might never have occurred to us yet resonate with us emotionally. The goal of the guitarist as artist is to come up with a sound of one's own, instantly recognizable as you. Consider the guitar heroes admired by millions: There is no mistaking Carlos Santana for Eddie Van Halen, or Jeff Beck for Jimmy Page. It is unlikely that you will confuse Wes Montgomery's sound with Jim Hall's or John Scofield's. Each of these players has carved out a sonic niche occupied by him and him alone. In the end, if you seek recognition as an artist, you, too, will need to find a tone that is your own.

That said, we all start by emulating our heroes, and many players are happy to be more craftsmen than artists, ready and eager to supply whatever sound a studio or live situation requires; there is an art to that as well.

This chapter should help quickly put you in the ballpark when you want to capture the signature sound of a particular style of music, as well as give you a little insight into the tone tricks of some famous guitarists. I have divided it up by musical categories; within each we will deal with general sounds and specific players. If you want the sounds to resemble the famous recordings, I recommend using some judicious Tube Compressor in the Rack section for each of the routings. And, as with learning to play music, it is terrific ear training to sit down with your favorite guitar records and attempt to match the tones.

Blues

Blues guitar covers a surprisingly large amount of tonal ground for music that is largely composed of three chords: from the raw Mississippi backwoods bar sound of a Junior Kimbrough to the sophistication of a B.B. King, and from the Texas shuffle of a Stevie Ray Vaughan to the jazz-influenced swing and funk of a Robben Ford. Here we will touch on a bit of each.

Raw Blues

The raw electric blues of R.L. Burnside or Pat Hare with Howlin' Wolf is all about lo-fi mystery.

- A great place to start is with the Tube Vintage Combo. Early blues artists often used small, inexpensive amps like the Silvertone upon which this is based. This amp also captures some of the tweed Fender tones that were popular at the time.
- Set the amp as in Figure 10.1. We won't be using the spring reverb, because most amps of the period lacked reverb. We will use routing Preset 1 to keep it simple.

Figure 10.1 Keep the amp reverb down and the volume up.

- The matching cabinet really enhances the lo-fi sound (see Figure 10.2). To bring out the big low end that characterizes this tone, we use the Velo 8 mic in combination with the Dynamic 57 for some presence. A fair amount of midsize room gives us the "bleed" that was common on those early blues recordings. Often one mic was used for the whole band!

Feel free to pan the mics in whatever fashion works. If the recording is just guitar, bass, and drums, you might want to spread the guitar sound, whereas if we are talking Muddy Waters' Chicago band with harmonica, piano, and another guitar as well, keeping it focused may be best.

Texas Blues

These days, when people think Texas blues, the first name that comes to mind is Stevie Ray Vaughan, and we'll get to him. But SRV is heir to a bunch of blues legends who also originated in Texas, including Clarence "Gatemouth" Brown, Albert Collins, and the father of the electric blues, T-Bone Walker. Lest we forget, Billy Gibbons of ZZ Top

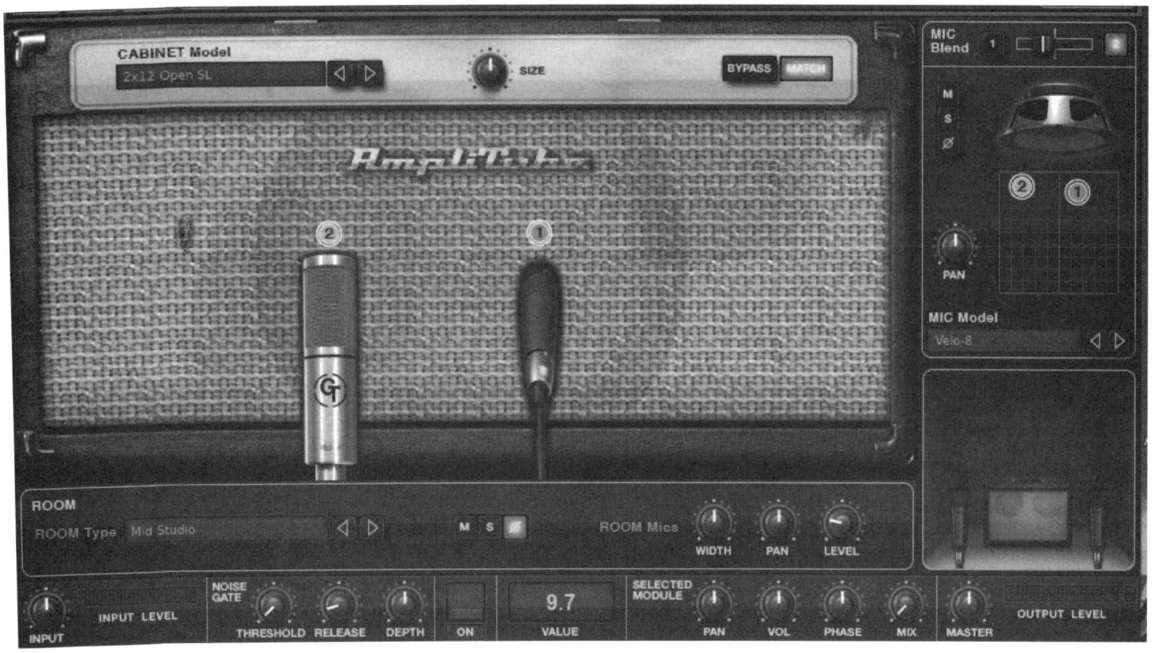

Figure 10.2 The Velo 8 helps bring out a smooth, warm low end.

put Texas guitar playing on the map well before Stevie, warping the blues into a pop sound that we will cover shortly.

T-Bone, Gatemouth, and the Iceman

Texans T-Bone Walker, Gatemouth Brown, and Albert Collins represent a major chunk of blues history not contained in Chicago, Mississippi, or the King family (B.B., Freddy, and Albert—no relation). All three of these swing/shuffle-oriented players used Fender Twin Reverbs or similar amps for much of their careers—the lack of coloration in its clean, warm, but present tone demonstrated just how much of their sound was in their fingers. I recommend the American Vintage T matched with the 2×12 Open SL to approximate the warmer tone of the '60s Fenders they favored.

- T-Bone's guitars often had pickups that were electronically out of phase, producing a unique honking tone. To approximate this with a guitar whose pickups are in phase, try setting the 7-Band Graphic stomp as shown in Figure 10.3 and using the bridge pickup in conjunction with the neck pickup. Adjust for single-coil or humbucking pickups and to taste. Keep the amp set clean and the tone present but not harsh.
- Gatemouth played with his fingers on the neck pickup of a Gibson Firebird sporting mini-humbuckers or P-90s. The sound should be fatter than single-coils but thinner than a full humbucker.
- Albert "the Iceman" Collins used his fingers as well, but his cool tone was the result of driving a Twin with the bridge pickup of his Fender Telecaster just hard enough to break up the slightest bit.

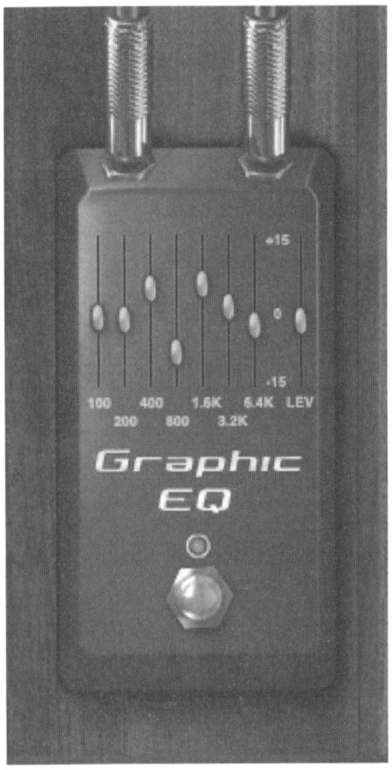

Figure 10.3 Set the Graphic like this to emulate the sound of two pickups electronically out of phase.

Stevie Ray Vaughan

It is important to realize that the lauded SRV tone is the product of a number of factors exclusive of effects, amps, and mic techniques. For starters, Vaughan normally sported extremely heavy-gauge strings on instruments set with a very high action. He tuned his guitars down to E♭, lessening the load a bit, but then proceeded to tear into them with both hands like a man possessed. Oh, and he also played at ear-splitting volumes. So no matter how close AmpliTube gets to reproducing his signature signal path, if you are timidly picking at a guitar with light strings and low action, you will get a weak approximation at best. Too, the Texas guitar slinger used various rigs at various stages. Here we will look at a simulation of his early equipment.

Vaughan's signal path began with the Ibanez Tube Screamer, which he helped make the go-to overdrive for this brand of rocking blues. Note how the Drive is set low and the Level high. The pedal is used as much to drive the amp into further distortion as for any overdrive it might produce on its own. Contrary to many other players, he preferred the wah to come after the overdrive (see Figure 10.4).

Stevie used two amps: one for dirt and one for clean, combining American and British tone in the process. A 40-watt Fender Vibroverb sporting a 15" speaker broke up nicely, with the speaker handling those heavy-stringed low-end riffs. A 100-watt Marshall Club and Country combo 2×12 added an edgy, clean sound for articulation. Here I have re-created the Vibroverb with an American Vintage B preamp linked to a 50W 6L6 power amp (see Figure 10.5), run through the 2×15 Closed B J130 cab (see Figure 10.6), and simulated the Marshall combo with a British Lead S100 (see Figure 10.7) matched with a 2×12 Closed Vintage (see Figure 10.8).

Figure 10.4 Note the minimal amount of Drive.

Figure 10.5 This setup approximates a Vibroverb. Note just a little spring reverb.

Feel free to adjust tones and panning to whatever works best for your situation.

Smooth Blues

Some like their blues raw and edgy; others prefer them smooth and sophisticated.

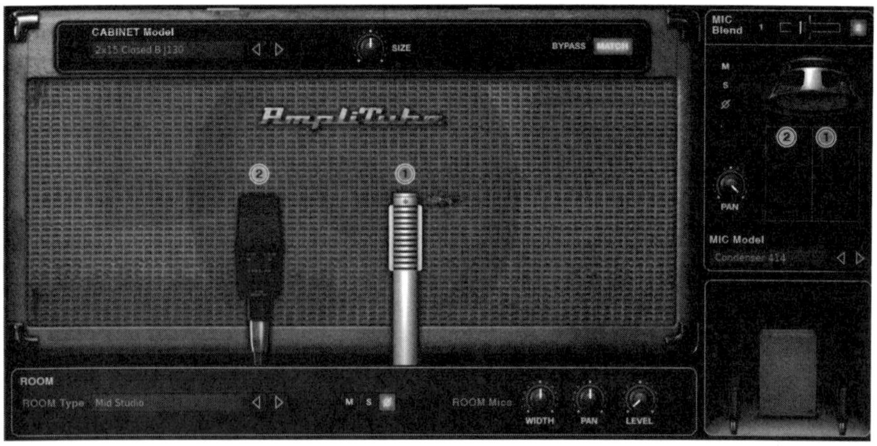

Figure 10.6 The 15" speakers will handle low-end riffs without mushing out.

Figure 10.7 This Marshall head is set relatively clean. Note no reverb.

Figure 10.8 This 2×12 should give the amp a combo feel.

B.B. King

Though he used Gibson's transistorized Lab Series amps for years, aside from some early records where he sounded like T-Bone, B.B. King exhibits a smooth, warm, tube-like tone. Play any humbucker-equipped guitar through the same clean rig as T-Bone and Gatemouth, leave lots of space in your phrasing, and you should be in B.B. Land.

Robben Ford

Robben Ford's brand of jazz-inflected blues and blues-infused jazz has attracted a following intrigued as much by his tone as by his playing. As with SRV, the tone cannot be entirely separated from his fingers and approach, but the all-but-unobtainable Dumble amplifiers that he uses are essentially Fender Twins on steroids, so with a little imagination we can come close in AmpliTube. Robben himself has been known to use a few pedals and a Fender Twin on gigs where he couldn't bring his Dumble.

Between switching his pickups from single-coil to humbucker, stepping on pedals, manipulating his volume, and taking advantage of the Dumble's gain stages, Robben is able to produce various tones including warm sustaining clean, low levels of grit, and screaming drive. His sound, however, always remains dynamic—never fuzzy or highly compressed.

- Begin with the same amp setup as his blues ancestors—the American Vintage T matched with the 2×12 Open SL. Be sure to set the gain and power section so that the amp breaks up only a little when hit with hard attacked chords (see Figure 10.9). This should ensure "give" and dynamics in clean mode. Just a hint of spring reverb will do.

Figure 10.9 Robben's clean tone.

- You may want to mic the cabinet with a Ribbon 121 and a Condenser 87 for some of Robben's warm, round tone. Add just a little Mid Studio Room for air, as in Figure 10.10.
- Robben often uses a TC Electronic 2290 Dynamic Digital Delay for a little slap echo. Here we use the Digital Delay in the Rack module set as shown in

Figure 10.11. Keep the mix low and roll off some highs. You don't want to hear the delay as such when you are playing; it should just add a nice tail to the note.

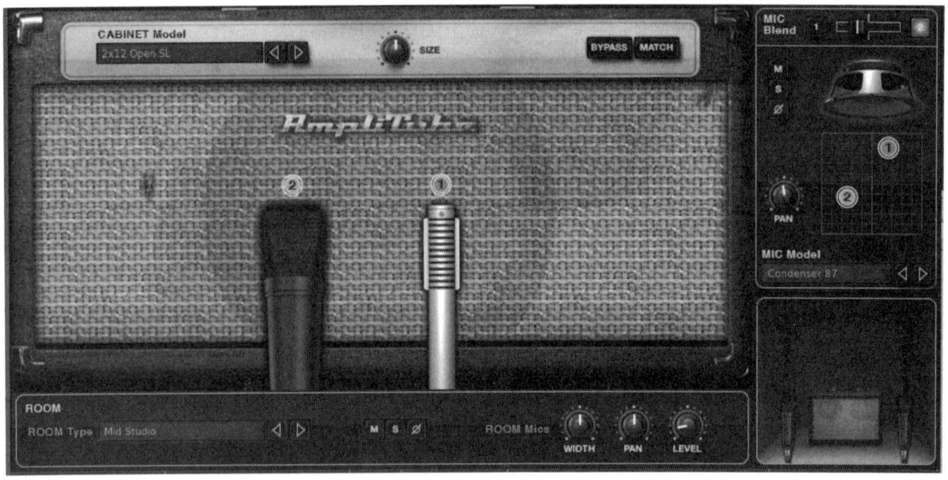

Figure 10.10 Ribbons and condensers for warmth.

Figure 10.11 An eighth-note delay should be plenty.

- To achieve the different levels of distortion at Ford's disposal, I recommend setting up three Diode Drives in the Stomp module, set as shown in Figure 10.12. Depending on how the amp reacts to your guitar pickups, you may want to use the drive on the far right as your clean sound to fill it out. The middle drive represents one level of medium dirt that can be found on Ford records. Adding the far-left pedal to the far-right one will emulate some of the California guitarist's earlier, more distorted tones.

Blues Rock

See the "Classic Rock" section later in this chapter.

Country

We're talking country that sounds like country here. If you're looking for "modern" country, you would do well to see the "Classic Rock" and "Pop" sections.

- For classic '50s and '60s country—Johnny Cash, Buck Owens, George Jones—all you need is the old standby, the American Vintage T, matched here with the 2×12

Figure 10.12 Driving one distortion with another helps retain dynamics.

Open T J120 for a little extra twang. A dab of spring reverb, the occasional Opto Tremolo, and EchoMan (to emulate a tape echo) are all the effects you will require. The Ribbon 121 and a Condenser 87 are a good place to start miking to take a little edge off the JBL speaker model.

- The '70s saw the odd phase shifter on records by Waylon Jennings, Merle Haggard, and Emmylou Harris, so throw a Phaze Nine into the Stomp section to re-create this era.
- The '80s were when pop elements began to creep into country. I struggled to get this era's sound with a chorus until I realized that they were using a pitch shifter set something like Figure 10.13. This also began the reign of the MXR Dyna Comp, so toss a Dcomp at the beginning of your pedal chain.

Figure 10.13 Note that the blend is set toward dry. The Fine tuning is set at 9.

The current cream of the country pickers, such as Vince Gill, Brad Paisley, and session ace Brent Mason, stick to this same basic rig. They might substitute a Fender Deluxe Reverb (American Clean 1) or a Vox AC30 (British Blue Tube 30TB), but they stay clean. The pedals may be boutique, but they still fall into the compressor, tremolo, delay, and reverb camps.

Rockabilly

Rockabilly is all about the slap echo. An EchoMan set to an eighth-note triplet is what you want, with only a few repeats at most. The difference between this and a blues, rock, pop, or country delay is that here you want to hear the repeats as such, not just as an appendage to the original note. You still don't want the repeats to get in the way of the original, so a setting like you see in Figure 10.14 is plenty.

Figure 10.14 A good rockabilly Delay setting.

Brian Setzer

The Stray Cats came out of nowhere in the '80s to revive the sound and look of rockabilly. Brian Setzer is a fully schooled guitarist who uses his knowledge of chords and scales to extend the genre's vocabulary.

- The Long Island born and raised guitarist favors Fender Bassman heads with their original 2×15 bottoms, so the American Vintage B is a good place to start. A Dynamic 57 and a Dynamic 421 will bring out all the high end the 15s have to offer.
- Setzer plays with more grit than your traditional rockabilly player, so I found that placing an OverScream with the drive way down and a Delay in front of the Vintage B (see Figure 10.15) got the tone pretty close.

Figure 10.15 The Delay blend is set dryer here, as the overdrive tends to bring out the effect.

Pop

One of the great things about pop music is that it allows for an endless number of potential rig setups to produce a nearly infinite array of tones. I will cover some classic pop tones in this section. If there are no current bands represented, it's because there is no telling which ones will still be around when the book comes out. Instead, I'll stick to bands that people still listen to after a decade or more. For the most part, modern pop bands are recycling these sounds anyway.

The Beatles

Vox jangle of the Beatles is easily attained thanks to AmpliTube's terrific British Copper 30TB emulation of the early Vox combos used by the Beatles and other British Invasion groups. A setting like Figure 10.16 will provide a perfect "I Feel Fine" clean sound when bridge and neck pickups are combined, while breaking up just the right amount for "Ticket to Ride" when just the bridge pickup is engaged (adjusted, of course, for single-coil or humbuckers).

Using these two pedals, set as shown in Figure 10.17, separately or together should put you in "Happiness Is a Warm Gun" and "Helter Skelter" territory.

The Edge

One of the signature sounds of post-Beatles pop is the delay-drenched chime of David Howell Evans's guitar. You might know him better as the Edge, or just Edge. His experiments with rhythms, set up by carefully worked out timings on his Electro-Harmonix Deluxe Memory Man, sent U2 on their way to becoming megastars and influenced generations of subsequent pop guitarists in bands such as Arcade

Figure 10.16 These settings should produce vintage Vox pop tones.

Figure 10.17 These pedals will help you get Beatle-esque distorted tones.

Fire; the Yeah, Yeah, Yeahs; and the Killers. The equipment to emulate his sound is all here in AmpliTube; what you will have to bring to the party is a steady right hand (if you play right-handed) capable of picking parts perfectly in time with the delays. The good news is that by picking at entirely reasonable tempos, you can sound as if you're playing a continual flurry of notes.

- If you want a—you'll pardon the pun—edgier sound, start with the British Copper 30TB set much like the Beatles tone. If you want access to a cleaner sound, use the British Blue Tube 30TB.

- Arrange a set of stomps like Figure 10.18. The compressor will help you keep those picked notes even, and the OverScream will add grit if you need it. Set the delay time for an eighth-note triplet.

Now play quarter notes at about 120 BPM, and you have instant Edge.

Figure 10.18 The eighth-note triplet setting will add swing and syncopation.

Rock

As with pop, rock can cover a wide range of tones. I have divided it up into classic and modern sounds.

Classic Rock

Classic rock tends to be more sonically homogeneous than modern rock, and—Hendrix excepted—it sticks to guitar, distortion and/or fuzz, and amp, with maybe the occasional wah pedal.

ZZ Top

You could file early ZZ Top under blues, but with *Eliminator* they fought their way free of that genre's restrictions into mainstream mega-sales, earning themselves a pop appellation. Billy Gibbons, one of the great guitarists in any genre, has purveyed a wide range of great guitar tones in his time, unified by his unerring rhythm, taste, and squealing pick harmonics. He is in this section because ZZ Top is a classic band, and there is no doubt that the Reverend Willie G rocks.

Whatever amps he uses, Gibbons likes to keep the sound dynamic, responsive to pick attack and able to clean up if need be. On both early and late ZZ Top recordings, Gibbons liked to double his rhythm parts either naturally, by overdubbing them, or artificially with delay and then hard pan them left and right. This allows him to use lots of bass on one or more of the rhythm parts without muddying up the center where the bass guitar is found. Therefore, any emulations of the sound of "the Lil' Ol' Band from Texas" should start with routing Preset 2.

Early ZZ Top. This rig represents Gibbons' tweed Fender gear on some of the initial records.

- In Signal Chain A, we have a Diode Overdrive set like Figure 10.19 and an American Tube Clean 2 amp through a 1×12 Open Vintage miked with a Dynamic 57 and a Condenser 87, panned hard right.
- In Chain B we have a Diode Overdrive set like Chain A and an American Tube Clean 1 through a 1×12 Open Modern miked with a Dynamic 57 and a Velo-8 to emphasize the low end, panned hard left.

Figure 10.19 Set the Diode Overdrive for a bit of boost and a little grit.

Late ZZ Top. Gibbons later switched to largely Marshall stacks and especially the Marshall rack-mount preamps. He also began adding a fuzz flavor to many of his rhythm parts and solos. This rig should get you close.

- In Signal Chain A we have a Diode Overdrive for milder dirt, as well as a gray Fuzz Age 2 fuzz set like Figure 10.20. These go into a British Tube Lead 1 driving a 4×12 Closed Vintage 1 miked with a Dynamic 57 and a Condenser 87, panned hard right.

Figure 10.20 Set the fuzz and boost like so.

- In Chain B we have a red Fuzz Age driving a British Tube Lead 2 into a 4×12 Closed Modern 2, miked with a Dynamic 57 and a Velo-8 to emphasize the low end, panned hard left. Rack B has a Parametric EQ with a 9.14-dB boost at 117.6 Hz to further goose the bottom.

Jimi Hendrix

AmpliTube 3 includes all of the amps and effects from the AmpliTube Jimi Hendrix package that used to be sold separately. In addition, it includes all of the presets from that series, matched to particular songs and parts of songs.

I reviewed AmpliTube Jimi Hendrix when it came out and was blown away by how accurately the sounds were reproduced. You may be surprised by how few of them involve large amounts of fuzzed-out gain. I did a recording for the review, and even though I have never made a study of playing like Hendrix, I was able to reproduce many of his recorded sounds just by dialing up one of the presets.

There is one tip I would like to impart here: After beating my brains out trying to get a certain slightly broken-up Hendrix rhythm tone out of a variety of amps, I discovered that the XS Fuzz with the guitar volume rolled off a bit delivered exactly what I was looking for.

Jeff Beck

It is said that there are great guitar players and then there is Jeff Beck. Unfortunately, no one has come up with a way to capture his inventiveness, whammy-bar dexterity, and sheer genius in a software application. That said, you can set up a rig that will produce Jeff Beck–like tones.

Beck went through a saturated period in the '80s but for the most part favors a medium amount of gain so that he can manipulate his sound with the guitar's volume knob. I recommend a British Lead S100 with a 50W EL34 power stage to emulate his JTM45. Diming out the volume and gain will produce enough distortion for some solo work when the instrument level is maxed but will still clean up when the guitar volume is lowered. Beck rolls the bass all the way off. Using a 4×12 Closed 25 C cabinet miked with a Dynamic 57 and a Ribbon 121 will keep the sound from getting too edgy.

A Diode Overdrive with the Distortion kept low will give you the kind of boost that Beck gets from the coveted Klon Centaur overdrive, kicking the amp up a notch without losing its character.

Beck has the bridge pickup of his Stratocaster wired to a tone control, and it is the delicate manipulation of its high-end roll-off that produces many of his new-millennium tones. I recommend that mod for Strat players regardless of the kind of music you play. Throw away your pick; practice conjuring up specific notes with the whammy bar, and you'll be on your way to the land of El-Becko.

Sixties British Rock

I have covered classic Clapton and Jimmy Page tones elsewhere in the book, but they were just the tip of the iceberg. England in the '60s was guitar-tone heaven: Free, Bad Company, Humble Pie, early Fleetwood Mac, Jethro Tull, the Bluesbreakers—everywhere you looked, there was a guitarist with a Les Paul grinding out this huge sound. The current country scene is recycling these terrific tones as we speak.

A simple rig containing a British Lead S100 set like Figure 10.21, through a 4×12 Closed 25c, driven by a Diode Drive with the Distortion at 0.39 and the Level at about 3.83, will get you right in the neighborhood. The rest will rely on your pick-hand attack and your finger vibrato.

Modern Rock

Modern rock is a flexible category (as are most), bleeding into pop, metal, classic rock, and even alt country. If we can agree that it should be primarily guitar based (preferably somewhat distorted), it still encompasses everything from Stone Temple Pilots to the Kings of Leon.

Following is a rig that should let you access most of the sounds of contemporary rock music.

Amp and Cabinet

- Vox amplifiers and their clones have caught up with—and some might say surpassed—Marshalls when it comes to usage for the modern rock sound. The British Blue Tube 30TB set like Figure 10.22 paired with a humbucker-equipped guitar should give you a nice, present, clean sound when the guitar volume is backed off

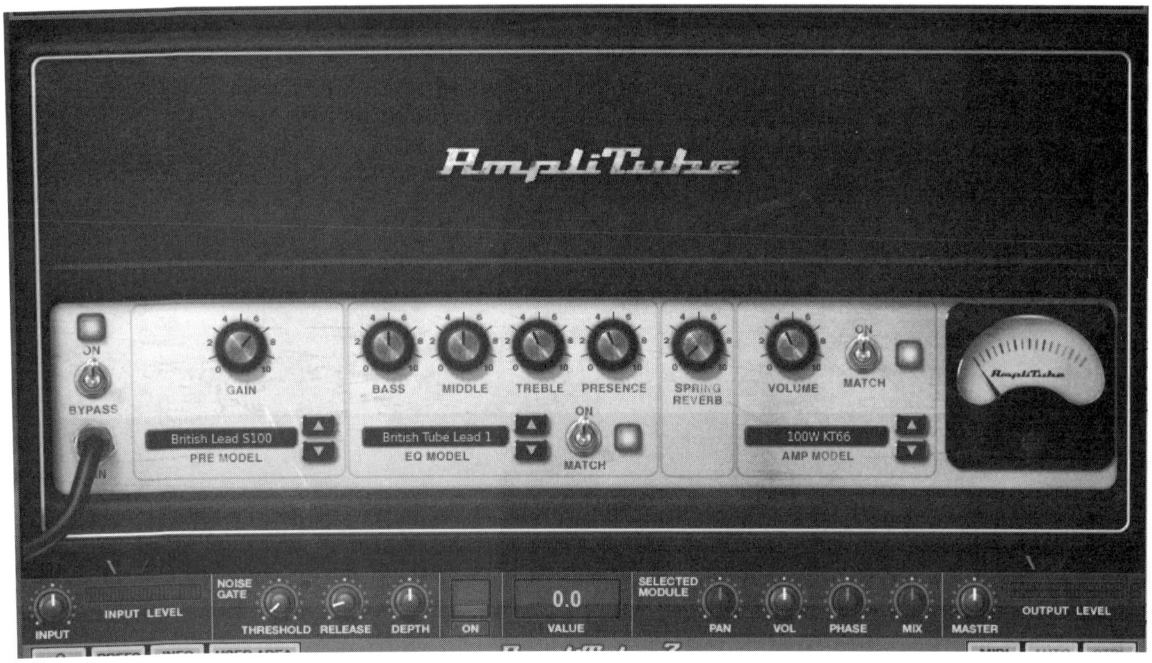

Figure 10.21 This amp setup combined with a Diode Overdrive will provide a classic rock tone.

Figure 10.22 This should give you a clean tone with plenty of chime.

and just a little grit when it is full up. For single-coil guitars, turn the volume up and maybe add some more Cut to the tone to keep it from getting thin or piercing.

- Pairing it with a 4×12 Closed 25 C cabinet will make the sound a little bigger.

Stomp Module

- In the Stomp section (see Figure 10.23), we start with a red Fuzz Age. Fuzz Face distortion has experienced a resurgence among players of the new millennium; its response to varying guitar volume input makes it very expressive. Full up, it will offer almost EBow-like sustain, especially when paired with the Diode Overdrive.

Figure 10.23 These effects all interact well together.

- A wah-wah is essential for rhythmic and filtering effects. The Wah 46 is an arbitrary choice; all the AmpliTube wahs sound great, so just pick the one that you like best.
- In addition to fattening up the Fuzz Age and adding sustain, the Diode Overdrive may be used alone to drive the amp into a little more crunch. Likewise, it can push the OverScream harder for more sustain when soloing.
- The OverScream pairs nicely with the British Blue Tube 30TB for those overdriven strums, crunch rhythms, and edgy double-stops. It, too, will respond to instrument volume level.
- The Whammy pedal has wormed its way into the world of must-have effects, whether for synth-style sweeps or steady octaves.
- Tremolo has, for the most part, replaced flanging, phasing, and chorus at the moment as the modulator of choice.

Rack Module

- We start the rack module (see Figure 10.24) with a Tube Compressor to add some warmth, even out the sound, and help pull it together. Go easy on squashing the attack and try to keep the gain level the same as when it is off.
- The Digital Delay can be engaged or not. Leave it and the Digital Reverb off for a real in-your-face sound or if you want to let the mixer handle all of the ambience. An eighth-note triplet used in a Left/Center/Right setting with the feedback at 33%, the Filter at 1830 Hz, and the Mix at 11% should spread your sound across the spectrum without muddying the mix.
- The Digital Reverb should really be considered an effect, used only when you want to sink your sound into a wash of ambience.

Figure 10.24 These ambient effects can add size and mystery.

Of course, you can add any other effects that you choose, but these basics should take you a long way in the world of modern rock.

Metal

Metal has many styles, genres, and sub-genres, from Death metal to Christian metal. What most share is a love for distortion—lots and lots of distortion. Early metal, such as Black Sabbath, Iron Maiden, and Judas Priest, used less than later bands, such as Pantera and Slipknot, but all use considerably more than the rock players of the '60s.

Eddie Van Halen

When Van Halen burst onto the Sunset Strip scene in the Los Angeles of the late '70s, the world discovered a guitarist who would revolutionize the instrument. Though Eddie Van Halen has influenced thousands of guitarists, few realize how little gain he initially employed, especially when compared to subsequent metal players. Though he increased it some in later years, the gorgeous sound and distinct articulation of his early recordings were the result of a relatively weak humbucker driving an amp that lacked the massive preamp gain of future metal amplifiers.

- We will use the Metal Lead V in his honor, but I recommend backing the Gain off from its default of 6.7 to 2.9. This should give you plenty of drive for chords while remaining articulate.
- Our trusty Diode Overdrive turns it into a two-channel amp and offers some boost for solos. A Metal Flanger will provide the patented whoosh of some Van Halen tunes.
- Setting up the Rack section as in Figure 10.25 with a quarter-note delay and a 3.67s decay on the reverb will give you the stadium sound. Leave the Pitch Shifter off for

early Eddie and turn it on in Split mode for his later tone. (Stereo seems to muddy the sound a bit.)

Figure 10.25 The Pitch Shifter will bring the Van Halen sound into the '80s and '90s.

Maiden and Priest

An easy way to start is with preset 80 metalz A, but the twin guitar attacks of these classic British Metal bands definitely calls for routing Preset 2. Call this up, and it will leave the 80 metalz A in Chain B. Add a British Tube Lead 1 to Chain A. Pan the Rack modules left and right. Add a Digital Delay to Rack B with a delay time of 15ms, the Mix all the way wet, the Filter all the way up, and the Feedback all the way down. This should give you a nice spread, with the amp reverbs adding the stadium sound.

Modern Metal

The gain kept increasing into the '90s with amps such as the Mesa Boogie Rectifier and the Randall Warhead pushing the envelope of nastiness when it came to distortion. AmpliTube offers a variety of these types of amplifiers, so finding this kind of sound should be easy. Just a couple of suggestions:

- As I mentioned in Chapter 8, "Effects," modern metal players get almost all of their gain from their amplifiers. Set the amp for the appropriate grind and then use only an overdrive of some sort for solos. Any kind of more extreme distortion pedal is likely to turn your tone to pure compressed fizz.
- Part of the modern metal sound is plenty of low-end thump and chunk. Be sure that the bass player's tone does not conflict with yours by dialing out the appropriate frequencies from the bass.

Jazz

Jazz guitar has come a long way from the universal sound of an archtop hollow-body equipped with a pickup through the cleanest amp available. Fusion introduced solid-body and semi-hollow guitars, rock tones, volume, and effects to this American art form. Post-modern players may be equally at home with clean be-bop or computer-enhanced free improvisation.

Classic

Electric guitar jazz originated on dark-sounding hollow-bodies, but you can get this historically warm, fat sound out of any instrument with a little help from the Parametric EQ.

- Start with the American Vintage D for the most clean headroom combined with tube warmth. The 1×12 Open Vintage will simulate a small jazz combo, or try the 2×12 Open SL for an early Twin Reverb "Wes" tone.
- Mike the cabinet with a Dynamic 57 and a Condenser 170 panned slightly left. Then move the two Mid Studio room mics a little closer and pan them hard right.
- Set the Parametric EQ to add 3.5 dB of Gain at 124 Hz, with the Q at 0.16. Roll off −13 dB of gain at 10 kHz with the Q set at 0.70.

This should result in a warm but not muddy classic jazz sound. As always, adjust the tone to your instrument and needs.

Modern

In the late '60s, musicians such as John McLaughlin and Larry Coryell combined rock attitude and volume with jazz harmony to create what would become known, for better or worse, as *fusion*. The genre got a bad rap due to an excess of athleticism at the expense of creativity on the part of some players and the morphing into "smooth jazz" of others, but there is nothing wrong with the basic concept.

The main problem from where I sit is that while serious jazz guitarists spend thousands of hours learning to play this complex and challenging music, they spend way too little time working on the tone issues and specific techniques required to play distorted guitar at higher volumes. Merely plugging any old distortion between your guitar and a clean amp will rarely suffice. Notable exceptions are players such as Scott Henderson, Allan Holdsworth, and Larry Carlton, who have obviously put the same kind of work into the sound as the notes. Here are some tips if you are a jazz guitarist seeking to add some rock sounds.

Distortion can be an expressive device. The trick is to not use too much, so that the dynamics of your picking still come through. For said dynamics to come into play, you must work on hitting the strings harder with your pick than is common in be-bop. Digging in will reveal an entirely different tone than playing softly. When you are using overdrive, both have their place.

Exercise 1

1. Open AmpliTube to the default setting.
2. Add an OverScream distortion with the Drive at 7.34 and the Level and Tone set at noon.
3. Pick some single notes lightly using the neck pickup.
4. Now hit the string as hard as you can for a few notes.
5. Listen to the difference in the amount of distortion.

Experiment with different amounts of Drive to see what works for you.

Delay can be another issue. When you are delivering a flurry of high-speed Coltrane licks, too much delay can turn the sound into a muddled mess.

- Matching the delay time to the tune's tempo can help it blend into the mix.
- Using the rack Digital Delay lets you roll off the highs with the Filter, letting you use a higher mix without stepping on the original signal.
- For the most part, if you can actually hear the repeats when the band is playing, you are using too much or too bright a delay.

Pat Metheny

Before he could legally drink in many places, Pat Metheny had helped revolutionize jazz guitar. Starting with the same big hollow-body as his idols, he ran it through delays and pitch shifting for a unique sound. Here's how you get it:

1. Start with the default AmpliTube setting and then switch to routing Preset 2.
2. Pan Cabinets A and B hard left and right.
3. In Rack A, install a Digital Delay set: Double, Delay 1/8d, Feedback 15%, Filter 363 Hz, Mix 7%.
4. In Rack B, install a Pitch Shifter set: Shift Coarse 0, Shift Fine 8, Split, Mix 100%. Install a Digital Delay set: Double, Delay Time 1/4, Feedback 11%, Filter 303 Hz, Mix 3%.

Note the extremely small mix percentages. You don't really want to be aware of the delays as independent notes, just as ambience and spread.

Allan Holdsworth

When it comes to fusion, few are as fleet-fingered as Allan Holdsworth, but it is his violin-like tone that makes his dense excursions so accessible. A unique chordal style is enhanced by a modulated sound that is lush yet crystal clear. We need two rigs to approximate his rhythm and lead tones.

Rhythm Tone

- To match Holdsworth's full sound, we need a dual rig featuring an American Vintage D in Path A driving a 2×12 Open T J120 and an American Tube Clean 2 in Path B pushing a 2×12 Open Vintage.
- A Compressor in both Stomp A and B slots will help bring out the modulation we will be installing in the racks.
- Like Pat Metheny, Holdsworth uses a variety of delays and modulation to spread his sound. Set one rack to look like Figure 10.26.

Figure 10.26 Remember to set the Pitch Shifter to Split.

- Set the other rack to look like Figure 10.27. The chorus will add richness to the chords.

Lead Tone. Holdsworth's solo sound is very difficult to nail, having much more to do with his touch than with his equipment: The man has sounded virtually the same playing half a dozen different amps and guitars. That said, the key is plenty of sustain with full tone, but clear, and with no fuzzy edges. It has a vocal quality that comes in part from his whammy-bar manipulations but also from the boosting of certain midrange frequencies.

- Setting up in stereo routing Preset 2, using almost any clean, warm tube-amp rig will work if combined with a pair of identical pedalboards set up like Figure 10.28.

Figure 10.27 Note that the delay is set for 1/8d as opposed to the other rack's 1/4.

Figure 10.28 Set these in Stomp A. Save each stomp setting as a preset and then recall them all in Stomp B.

- Set up the Pitch Shifter in Rack A like the rhythm rig but back the mix down to 11% and don't use any chorus.

Now go and practice—a lot!

Summary Questions

You can find the answers to the Summary Questions in the Appendix at the end of this book.

1. What is a good amp for raw vintage blues?
2. Which microphones combine well with this sound?
3. Which stomp effect can help approximate the sound of two pickups electronically out of phase?
4. When going for a Stevie Ray Vaughan sound, the OverScream Drive control should be set:

   A. Low

   B. Medium

   C. High
5. Which mics will help achieve Robben Ford's smooth sound?
6. Which era of country music saw phasers being used?
7. Which effect was introduced to country in the '80s?
8. What delay length is emblematic of rockabilly?
9. Which amp will give you the most accurate Beatles tone?
10. What length notes do you play against the eighth-note delay to get the Edge's sound?
11. Which stomp is good for early ZZ Top tone? For Billy Gibbons' later sound?
12. Which fuzz is excellent for a Hendrix rhythm tone?
13. What is the essential '60s rock rig?
14. True or false: For a great Van Halen tone, use the Metal Lead V in its default setting.
15. Which should you use for a Pat Metheny jazz tone: a rack chorus or a rack pitch shifter?

11 Putting It All Together

Now that we've covered all of the elements in AmpliTube that help you record great guitar tracks, let's actually record some. In this chapter, we'll put into play many of the principles you learned in the previous chapters while recording using AmpliTube's 4-track recorder. If you have a well-set-up guitar, the full version of AmpliTube, a computer, and a quality audio interface, you already have everything you need to put together a great-sounding piece of music.

Let me take you through my steps in producing an instrumental tune using only the 4-track recorder, AmpliTube 3, an IK StealthPedal, and an iMac. First, I'll pull out my trusty Telecaster—okay, now it's time to make some music.

Step 1

Here we'll deal with the interface.

1. To begin with, I make sure I have the current drivers for the StealthPedal installed on the Mac.
2. Next, I plug the pedal into the computer with the supplied USB cord.
3. Going to the bar at the top of the window, I click on Settings, which reveals the Audio MIDI Setup option.
4. Clicking on the Audio MIDI Setup option reveals the Hardware Settings window (see Figure 11.1). I choose StealthPedal for my audio input, as that's where I'll be plugging in my instrument.
5. For my output, I choose an M-Audio FireWire 1814 interface that's hooked up to my Roland monitor speakers. I could also choose the StealthPedal as my audio out and monitor through headphones or powered speakers plugged into the pedal. For lowest latency, using the audio outs of the StealthPedal would be optimal, but I find that the latency through the M-Audio device is workable, so I continue to Step 2.

Step 2

We have four tracks to work with, and I want to dedicate three tracks to guitar, so bass and drums need to go on one track.

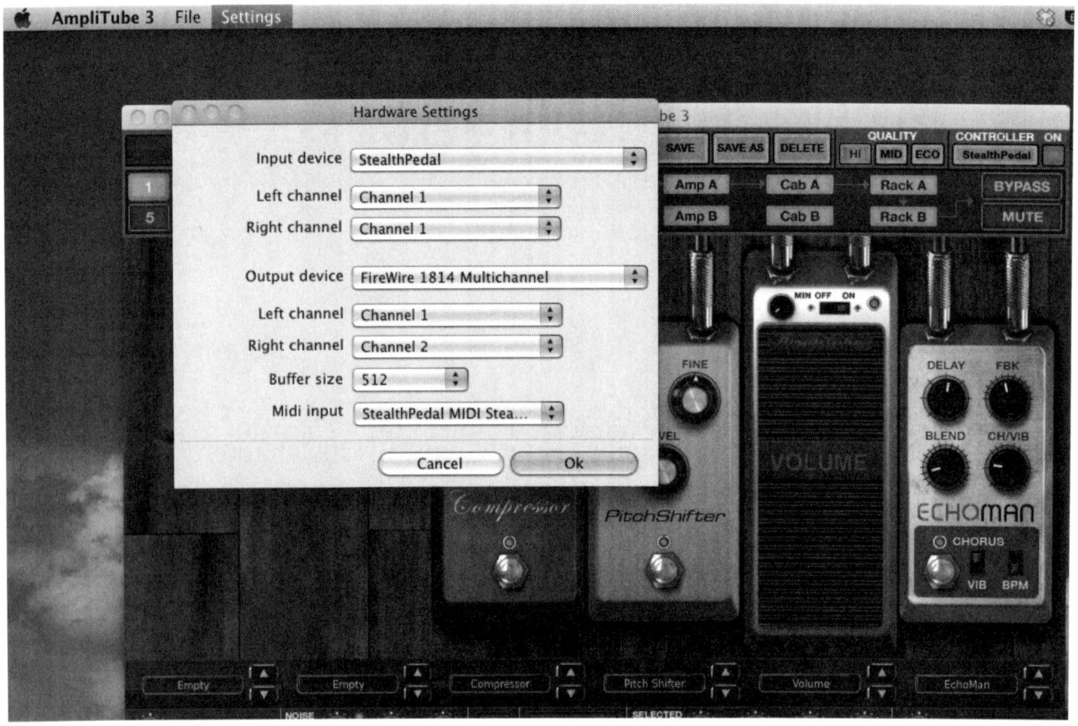

Figure 11.1 Here are the hardware settings that I used.

1. You could mike up your bass player and drummer and run them through a mixer into the StealthPedal, but I chose to put a bass and drum part together in Ableton Live and then bounce it to a stereo AIFF track that I saved to my desktop.
2. Alternatively, IK supplies a drum track with AmpliTube to get you started, or you could download one from the Web and add your own bass part using an octave effect and an AmpliTube bass amp. You can then mix those two tracks together, export them, and reload them into one track, leaving three tracks free for guitars.
3. For me, it is a matter of dragging the track that I created in Live into Track 4. If you use the AmpliTube file, your exported mix, or any download, it will be equally easy—just drag and drop.
4. I then add AmpliTube to that track. Shutting off the amp and cabinet modules, I use the Rack section's Parametric EQ and Tube Compressor to beef up the bass and drums (see Figure 11.2).

Step 3

Now we are ready to start recording guitar.

1. I have only three tracks left, but I hear four different guitar sounds in my head, so for the first recording action I set up AmpliTube on Track 1 using Preset 2 in the Routing section.

Figure 11.2 Adding EQ and compression to the bass and drums enhances the rhythm tracks.

2. For Amp A, I leave the default American Tube Clean 1 with its matching cabinet.
3. In the Stomp A section, I install a Big Pig with the Sustain rolled back as in Figure 11.3.

Figure 11.3 The Big Pig's Sustain is rolled back to keep this part defined.

4. The cabinet mics (a Dynamic 57 and a Condenser 87) are both panned fairly hard right but not identically. A little variance gives the right-side track a bit more width.
5. A Parametric EQ is used in the Rack section to roll off some low end, so as not to compete with the bass.
6. For the B routing, I choose a British Blue Tube 30TB for the amp, driven by a Diode Overdrive. Both pedal and amp are set bright—see Figures 11.4 and 11.5.
7. This amp's mics are panned left.

Figure 11.4 The Diode Overdrive's tone is set on the bright side.

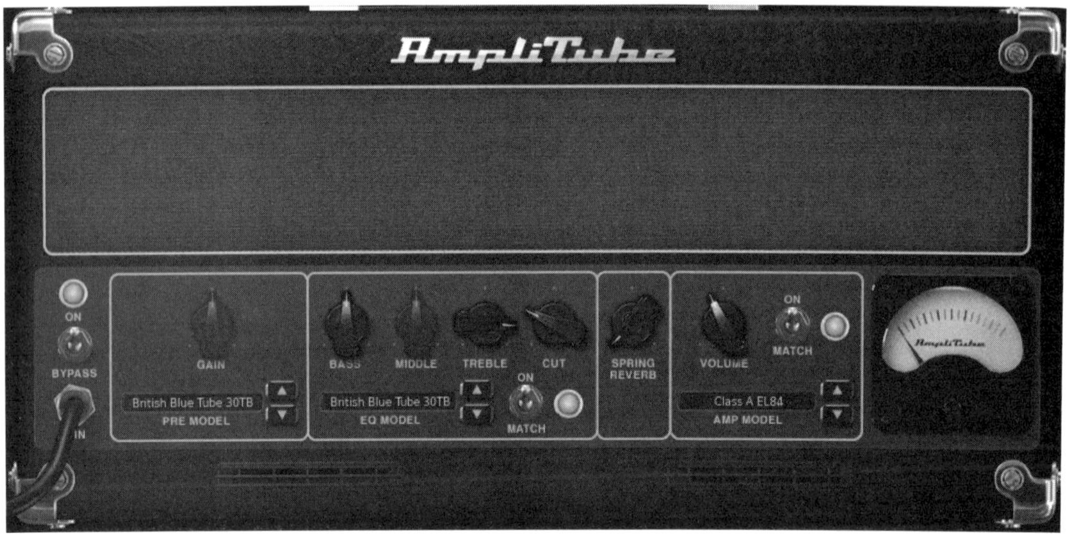

Figure 11.5 The Treble up and the Treble Cut down help set this left-panned amp apart from the fuzz sound panned right.

8. Another Parametric EQ is used in the B Rack section to roll off low end to prevent competition with the bass.
9. Now I'm ready to simultaneously lay down a pair of distinctly different-sounding power chords on a single channel—one a Vox-type tone on the left and one a fuzzed-out Fender on the right. I come down on Beat 1 and let the chord ring for two bars before striking it again, continuing this pattern to the end.

Step 4

For Track 2, I hear a chiming picked part.

1. This requires only one amp (a British Lead S100 set clean), so routing Preset 1 works fine.
2. A Compressor helps bring the ring out of a Pitch Shifter. The Shifter is set for 3 cents and blended into the background with the Level knob. An Echoman adds some further chime as well as some delay to give the part still more shimmer and to set it back from the power chords, as well as the eventual solo part (see Figure 11.6).

Figure 11.6 This group of effects is set to make a cool chiming sound for picking parts.

3. This part gets panned a little to the left in the recorder's panning section (see Figure 11.7) to leave the middle open for the solo.

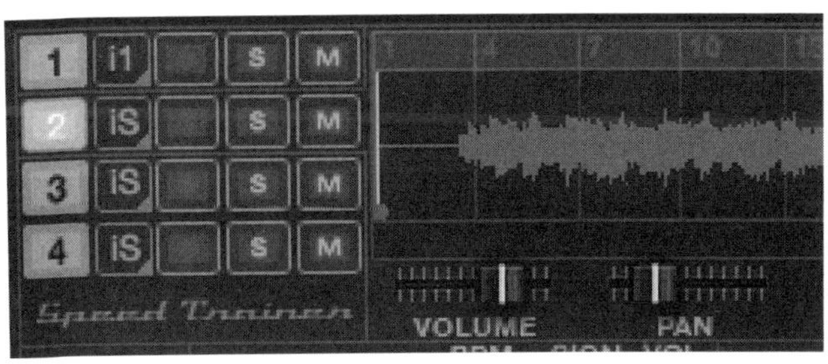

Figure 11.7 It is easiest to pan a mono track by moving the slider in the 4-track recorder's panning section.

4. A Rack Parametric EQ once again rolls off some low end and also some mids, so as not to compete with the snare and solo. A Tube Compressor keeps the part and its delays audible at low volume.

Step 5

Finally, it is time to go for the melody and the solo.

1. Using Routing Preset 1 again, I load one of my favorite AmpliTube emulations, the British OR, into Track 3. Set like Figure 11.8, it requires only a slight shove from a Diode Overdrive to sing as sweetly as Aretha Franklin.

Figure 11.8 This emulation of an Orange amp exhibits an attractive combination of grit and warmth.

2. A Ribbon 121 combined with a Condenser 87 smoothes out any remaining harshness, but, as this part is the main attraction, the sound is not yet compelling enough. It remains to the Rack section to put the cherry on the sundae (see Figure 11.9).

Figure 11.9 Some judicious Rack effects can make the difference between a lifeless and a lively sound.

3. There I use a Graphic EQ to roll off competing low end and take out some mids to let the snare poke through. The Tube Compressor helps even out any sloppy playing, as well as adding more warmth.
4. The first Digital Delay spreads the solo sound a bit. It takes a delay time of only 1/32 note spread left, center, right by the LCR setting to enlarge it significantly.
5. The second delay, set for 1/4 note and placed in the background by lowering the Mix and Filter, sits the solo in the mix, while adding a subliminal tail to the notes.

That's All She Wrote

There you have it—a complete musical piece recorded entirely in AmpliTube. Once I have everything recorded and balanced to my satisfaction, I need only hit the Export button in the GUI's lower-right corner. The Export File screen will appear (see Figure 11.10), asking me which file format (WAVE, AIFF, CAF, or SD2) and which bit rate I wish the file to be (16-, 24-, 32-bit float, 64-bit float) and where I wish to store it.

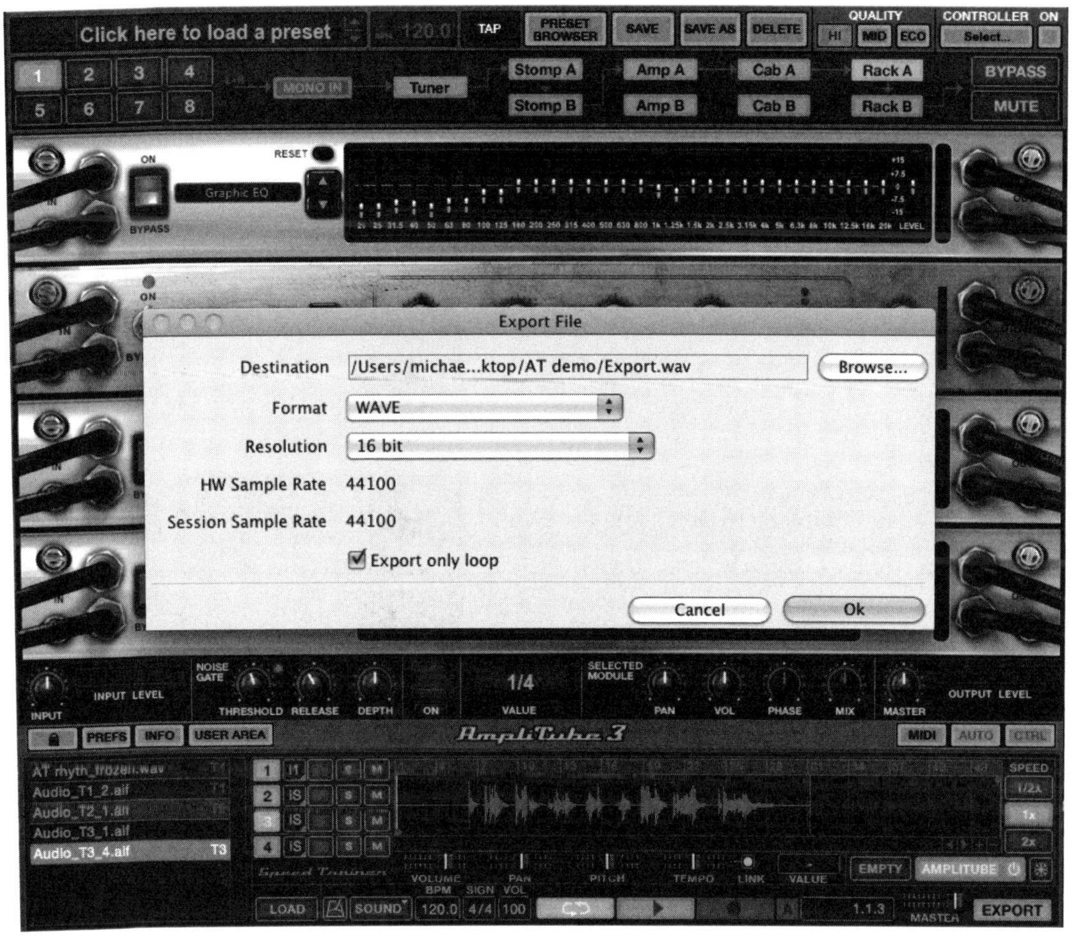

Figure 11.10 Your final product can be saved and shared.

Hopefully, you will be happy with the sounds that you record (inasmuch as guitarists are ever happy with their sound). Keep in mind that this is but the first step: AmpliTube offers the opportunity to spend countless hours experimenting with guitar tones, alone and in combination with other tones. Trust your ears and your creative spirit—and most of all, have fun.

Appendix: Answers to Summary Questions

Chapter 1

1. Which has a "woodier" tone: a solid-body guitar or a hollow-body guitar?

   A hollow-body guitar generally has a woodier tone than a solid-body guitar.

2. What is the difference between a hum-canceling pickup and a humbucker pickup?

   A hum-canceling pickup is any pickup that cancels 60-cycle hum, regardless of its sound. A humbucker is a specific pickup with two single-coil pickups wired next to each other in series.

3. What are two reasons to use new strings when recording?

   New strings will be more accurately in tune and will be brighter sounding.

4. What is the term for the height of the strings off the fretboard?

   This is called the *action*.

5. What is the difference between getting in tune and staying in tune?

   Getting in tune means that the instrument is in tune all the way up the neck and in every position to begin with. Staying in tune means that it remains that way as you play.

6. What should you do to your strings to ensure that they stay in tune when they are played and bent?

   You should stretch them by pulling on them.

Chapter 2

1. What is an audio interface?

   An audio interface is a device that takes the audio signal of your instrument, converts it from analog to digital, sends it to your computer, and then takes the digital signal coming out of the computer and converts it back to analog to come out of the speakers.

2. What is latency?

   Latency is the time that elapses between the moment that you strike the strings and when you hear the sound come out of the speaker.

3. What is a buffer?

   The buffer in your computer is where data is stored when it's retrieved but before it's used. Higher buffer settings allow more complicated processing; processing lower buffer settings is much faster, resulting in lower latency.

4. Where do I find the buffer settings in AmpliTube?

   In standalone mode they are in the Audio/MIDI Setup section. When you are using AmpliTube as a plug-in, they will be controlled by your DAW's settings.

5. What is AmpliTube's Quality mode?

   You can choose three levels of audio quality in AmpliTube. Higher levels require more CPU power.

6. Name three ways that you can deal with latency issues.

   Any of these is correct:

   - Reduce the Quality mode.
   - Reduce the buffer size.
   - Freeze tracks.
   - Record guitars early in the project.

7. What are the advantages of USB?

   USB interfaces are cheaper, and most computers have plenty of USB ports.

8. What are the advantages of FireWire?

   FireWire interfaces are faster than USB interfaces, helping to reduce latency and making them better able to handle multiple-track recordings.

9. What is direct monitoring?

   This lets you hear a direct signal with no latency when recording by bypassing the computer entirely, going straight from the interface input to its output.

Chapter 3

1. What does MIDI stand for?

   Musical Instrument Digital Interface.

2. What does CC stand for?

   Continuous controller. This is a knob, a slider, or an expression pedal that sends a continuous stream of 127 messages as it is manipulated from one extreme position to the other.

3. You can map a single controller to:

   A. One parameter

   B. Two parameters

   C. Any number of parameters

   Answer: C

4. Reversing the minimum and maximum percentages of a setting on the MIDI Control page has what effect?

   Turning a knob will have the opposite effect: in other words, if the knob was raising the volume when turned clockwise, it will now lower it.

5. How do you make one effect turn off and another turn on using one switch?

   Map them both to the same switch and set them for opposite minimum and maximum percentages on the MIDI Control page.

6. Give two reasons for using a manual controller with AmpliTube.

   It allows you to manipulate AmpliTube controls in different modules without having to switch screens. It permits you to change many parameters at once.

7. Which effects require an expression pedal?

   For maximum benefit, the volume, wah, and whammy effects need an expression pedal.

8. On which effects can you use an expression pedal?

   You can adjust the parameters on any effect using an expression pedal.

9. Why is a beat-oriented pad controller less than ideal for AmpliTube?

   Though the pads can act as switches, the units tend to be short on knobs and faders.

Chapter 4

1. What are two advantages to monitoring through headphones?

   Any of the following:

   - It's ideal for checking stereo panning, hearing unwanted noise, and tending to sonic detail.
   - It helps to shut out distractions.

- It allows you to work late at night without annoying others.
- It eliminates anomalies such as boominess and excessive liveness or deadness of rooms.
- It's comparatively cheaper than speakers.

2. What are two disadvantages to monitoring through headphones?

   It does not let you know how a room will affect the sound. Also, panning tends to sound more exaggerated through headphones.

3. What are two advantages to monitoring through speakers?

   It gives you a good sense of what the music sounds like in a room. Also, it usually provides better, more accurate bass information and EQ information in general.

4. What are two disadvantages to monitoring through speakers?

   Any two of these:

   - It can be annoying to neighbors, family, and friends.
   - Clicks, pops, string squeaks, and amp hiss are not as easily detected as with headphones.
   - Speakers are generally more expensive than headphones.

5. What are two things you should do when purchasing your monitoring system?

   Listen to them playing a musical source that features a wide frequency spectrum. Also, listen to a high-quality (CD, not MP3) musical source that you know extremely well and have listened to under many circumstances.

6. What is the most important thing to remember about a monitoring system?

   Familiarity is the most important factor in producing great tones that transfer well to other systems.

Chapter 5

1. What is the most important secret to getting great guitar tone?

   Context. Tone is relative to what is happening sonically around it.

2. What is the difference between American and British tone?

   American tone has a smooth warmth, resulting a big low end and an emphasis on the low mids. British tone conjures up an edgier sound with a hint of metallic clang produced by strong upper mids and a sharper high end, as well as a tighter bottom.

3. When does the power tube section of an amp have the greatest effect on the tone?

A. At high volume?

B. At low volume?

Answer: B. At high volume is when the power tubes have the most effect.

4. What is the tonal difference between Fender's Twin and Dual Showman amps?

   The Twin sounds more focused in the midrange.

5. Which amp models emulate the sound on the first Led Zeppelin record?

   Tube Vintage Combo.

6. Which model is best for the sound of bands such as Korn, Limp Bizkit, and Linkin Park?

   Modern Tube Lead.

7. Which set of tubes produces a more vintage sound in a Triple Rectifier model: 6L6s or EL34s?

   EL34s.

8. Which amp's control plate features a Sputnik icon?

   Orange.

9. How do you create Eric Clapton's Bluesbreakers Marshall combo?

   Start with a British Lead S100, substitute a 50W 6L6 power section for the British Lead's 100W KT66s, and switch the cabinet to a 2 × 12 Closed Vintage.

10. Which amp emulates Marshall's Master Volume series?

    Vintage Metal Lead.

11. How do you construct a Music Man amp?

    Exchange the Custom Solid State Clean's solid-state preamp for a tube one.

Chapter 6

1. What is one reason why you would want to use a 6″ speaker cabinet?

   To fit a guitar part in the mix without obscuring any of the other parts, the bass, or the vocals.

2. True or false: There is very little difference between the sound of the same speaker close miked in a 1 × 12″ cabinet and a 4 × 12″ cabinet.

   True. If you aren't picking up the other speakers with a room mic, they will sound very similar.

3. What sort of sound would be best produced by a 25-watt Celestion-equipped cabinet?

   Warm midrange.

4. What sort of sound would be best produced by a 120-watt JBL-equipped cabinet?

   Cutting high end.

5. True or false: Any speaker cabinet can sound good with any amp configuration.

   False. Although many speaker configurations will work with a particular head, some just won't.

6. True or false: With the same speakers loaded, a miked straight 4 × 12″ Marshall cabinet will sound very similar to a slanted version of the same cabinet.

   True. The only difference is in live application for room dispersal and balance in a stack configuration.

7. What are two styles of playing that go well with a 2 × 15″ cabinet?

   Surf guitar and Texas blues, among others.

8. What type of music favors the Celestion G-12-75 speakers?

   Post-'60s rock.

Chapter 7

1. To significantly change the sound of a recorded guitar, you must move the mic a minimum of:

   A. Two feet

   B. One foot

   C. One inch

   Answer: C. Moving a mic one inch is all it takes to significantly change the sound.

2. What is the result of moving a mic closer to the cabinet?

   More definition, increased highs and lows, and less room sound.

3. What is the result of moving a mic closer to the cone of the cabinet?

   More highs and better transient detail but fewer lows.

4. What are the three types of microphones commonly used to mike guitar cabinets?

   Dynamic, condenser, and ribbon.

5. What is proximity effect?

   The mic produces more low end when placed closer to the source.

6. Which dynamic mic does *not* boost the midrange frequencies?

   The Dynamic 441.

7. Which ribbon mic has become very popular for close-miking electric guitar?

   The Royer R-121, modeled here as the Ribbon 121.

8. What do you need to be careful about when using two or more mics?

   Phasing. Mics that are not properly in phase can thin out the sound in unpleasant ways.

9. Which quality modes allow room miking in AmpliTube?

   Hi and Mid modes.

10. How does panning work?

    Panning knobs set the volume of sound coming out of each speaker.

Chapter 8

1. What is the main rule when using effects?

   Less is more (except occasionally when more is more).

2. What is the definition of the term *dynamic range*?

   Dynamic range is the distance between the softest sounds produced and the loudest.

3. How does a compressor work?

   A compressor is a variable gain amplifier that acts as an automatic volume pedal. The compressor keeps the volume sent to the amplifier at a preset level. When volume sent to the amp drops, the compressor's amplifier keeps raising the volume in an attempt to maintain the same level as the initial attack.

4. Why is it called a "graphic" equalizer?

   The boost/cut sliders offer a graphic representation of the shape of the EQ.

5. What is the difference between a high-pass, a low-pass, and a band-pass filter?

   A high-pass filter lets the sound only above a selected frequency through, allowing just the treble to pass. A low-pass filter lets only the bass through. A band-pass filter lets you choose the frequency to be let through.

6. What is the difference between an envelope, an LFO, and a step filter?

   An envelope filter responds to the level of signal at the input, a low-frequency oscillator controls an LFO filter, and a step filter allows you to program the tonal shift in steps.

7. What is the difference between boost, overdrive, distortion, and fuzz?

   A boost pedal raises the volume level of the signal without adding distortion. An overdrive pedal delivers a smooth sound and a minimal amount of distortion, and it sounds the most like a real amp overdriving, cleaning up nicely when the guitar volume is backed off. A distortion pedal offers more gain and an edgier sound and may not clean up when you lower your guitar output. A fuzz pedal produces a fizzier sound and less focused distortion, and it may or may not add an upper-octave harmonic or clean up when the guitar volume is lowered.

8. Where is an Octa-V's upper octave most noticeable?

   When playing on the upper frets of the guitar, especially the highest three strings.

9. How does a pitch shifter differ from a harmonizer?

   A pitch shifter adds an unchanging interval, whereas a harmonizer can be set to change intervals depending on the key.

10. What is the difference between chorus, flanging, and phasing?

    A chorus effect uses an LFO to continually shift the pitch a few cents above and below the dry one by changing a short delay time. In a flanging pedal, a signal is delayed by a number of delays in the 5–10ms (millisecond) range, modulated by the LFO and fed back on itself, creating the signature whooshing. A phaser shifts one signal out of phase with the original. The second signal is treated with all-pass filters, each frequency with a different short delay (10–30ms). An LFO alters the filter frequency ranges over time, and this continually shifting phase creates the swooshing sound.

11. What is the difference between graphic and parametric EQ?

    Graphic is used to shape the overall signal, whereas parametric hones in on very specific frequencies.

Chapter 9

1. What is the difference between series and parallel routing?

   Series means that the signal is going out of one source directly into another source. Parallel means that two signals are traveling alongside one another.

2. Name two reasons to use routing Preset 1.

   It creates a focused sound that is easily panned into a specific location in the stereo spectrum. This is a good preset to create a specific vintage sound and/or when you want to use a single recognizable amp sound in mono.

3. Which routing presets allow parallel effects routing?

   2, 5, 6, 7, 8.

4. If you just want to have two cabinets, which preset would you use?

   Preset 3.

5. Which preset lets you use two separate amp and cabinet rigs in parallel, but still use all the effects in series?

   Preset 4.

6. If you wish to run parallel stomps with two separate rigs and series racks, which preset would you choose?

   Preset 7.

Chapter 10

1. What is a good amp for raw vintage blues?

   The Tube Vintage Combo.

2. Which microphones combine well with this sound?

   Dynamic 57 and Velo-8.

3. Which stomp effect can help approximate the sound of two pickups electronically out of phase?

   The 7-Band Graphic.

4. When going for a Stevie Ray Vaughan sound, the OverScream Drive control should be set:

   A. Low

   B. Medium

   C. High

   Answer: A

5. Which mics will help achieve Robben Ford's smooth sound?

   A Ribbon 121 and a Condenser 87.

6. Which era of country music saw phasers being used?

   The '70s.

7. Which effect was introduced to country in the '80s?

   Pitch shifting.

8. What delay length is emblematic of rockabilly?

   The eighth-note triplet.

9. Which amp will give you the most accurate Beatles tone?

   British Copper 30TB.

10. What length of notes do you play against the eighth-note delay to get the Edge's sound?

    Quarter notes.

11. Which stomp is good for early ZZ Top tone? For Billy Gibbons' later sound?

    Early: Diode Overdrive. Late: red Fuzz Age.

12. Which fuzz is excellent for a Hendrix rhythm tone?

    XS Fuzz.

13. What is the essential '60s rock rig?

    British Lead S100, 4×12 Closed 25c, Diode Drive.

14. True or false: For a great Van Halen tone, use the Metal Lead V in its default setting.

    False. For more accurate Van Halen tone, you should back the Gain off considerably.

15. Which should you use for a Pat Metheny jazz tone: a rack chorus or a rack pitch shifter?

    A pitch shifter will be closer to the subtle modulation of Metheny's early tone.

Glossary

2×12, 4×12, etc. This refers to the number, followed by the size of the speakers in an amplifier cabinet. For example, 2×12 means that there are two 12″ speakers in the cabinet.

action. This refers to the height of the guitar or bass strings off the fretboard.

AD/DA. This refers to audio-to-digital conversion and digital-to-audio conversion.

CC. This stands for continuous controller. A knob, slider, or expression pedal can send a continuous stream of 127 messages to control software actions.

DAW. This acronym comes from the term *digital audio workstation.* It refers to software recording systems such as Pro Tools, Cubase, Logic, Ableton Live, and so on.

distortion. When the top of a sound wave is flattened or "clipped," it is said to have become distorted. Our ears hear distortion as everything from analog warmth to nasty digital fuzz.

dynamic range. The distance between the softest sounds produced and the loudest ones.

FireWire. FireWire is Apple's name for the IEEE 1394 high-speed serial bus. It is a high-speed transfer system between your audio input/output and the computer.

GIGO. *Garbage In, Garbage Out* is a term used since the early days of computers to indicate that whatever you put into the system—code, input data—will be reflected in what comes out. If the code or data is faulty, the results will be faulty.

humbucker. This generally refers to pickups containing two coils side by side, wired in series and reverse polarity to eliminate 60-cycle hum.

hum canceling. This refers to any pickup that cancels 60-cycle hum, regardless of sound.

latency. This refers to the delay time between when audio or MIDI information is input into the software and when the results are heard coming out of the speakers.

MIDI. Musical Instrument Digital Interface, or MIDI, is an industry-standard protocol that enables electronic musical hardware (synthesizers, drum machines, effects), computers, and other electronic equipment (MIDI controllers, soundcards, samplers) to communicate and synchronize with each other and with software in the computer.

panning. Panning sets the volume of a particular sound coming out of each speaker—that is, if it is louder coming out of the left speaker, it is considered to be panned left.

parallel. This describes two signal paths running side by side.

power amp. This refers to the part of a combo or amp head that does the final level increase to drive the speakers.

preamp. This refers to the part of a combo or amp head that increases the instrument signal enough to drive the power amp.

proximity effect. This describes the effect of a mic producing more low end when placed closer to the source.

series. This refers to a signal path where the signal goes in one straight line.

tremolo. This describes the rhythmic fluctuation of volume from louder to softer and back.

USB. Universal Serial Bus (USB) is a common data transfer system between devices and computers.

vibrato. This describes the rhythmic fluctuation of pitch.

Index

Numbers

A

B

I

J

K

L

M